SpringerBriefs in Applied Sciences and Technology

For further volumes:
http://www.springer.com/series/8884

Lakshmi Kantha

Migration on Wings

Aerodynamics and Energetics

 Springer

Lakshmi Kantha
Aerospace Engineering Sciences
University of Colorado,
Boulder, CO, USA

ISSN 2191-530X   e-ISSN 2191-5318
ISBN 978-3-642-27924-9   e-ISBN 978-3-642-27925-6
DOI 10.1007/978-3-642-27925-6
Springer Heidelberg New York Dordrecht London

Library of Congress Control Number: 2012930488

Printed on acid-free paper

Springer is part of Springer Science+Business Media (www.springer.com)

Abstract

Birds have been an object of human emotions, admiration, fascination, inspiration, and, of course, envy. Humans have always marveled at their ability to take to the skies and in some cases, seemingly effortlessly carry out seasonal migrations over unimaginably long distances. Birds are an engineer's dream. Millions of years of evolution have perfected these flying machines. Their extremely light weight bone structure is essential to keeping them airborne. Their wings generate both thrust and lift, a feat that human-built flying contraptions had great difficulty imitating. They are not only ideally suited to flapping and in many cases soaring flight, but can be folded and stowed away into a compact package after the flight. Their flight muscles are adept at converting body fat directly into mechanical energy for flight. The eggs they use for propagation of their species are themselves an engineering marvel: extremely light weight but incredibly strong shell structures that enable the chicks inside to develop in safety. They are constructed such that they contain the nutrients necessary without leakage, yet permit diffusion of oxygen into the egg. Birds have also been the inspiration for human's successful effort to build heavier-than-air flying machines that have revolutionized the way we travel. But by far, the most fascinating aspect of birds is their ability to migrate nonstop very long distances to their breeding grounds during summer, and to warmer regions during winter. Only recently have man-made jet planes been able to fly similarly long distances without refueling. This book is an effort to explore the technical aspects associated with bird flight and migration on wings. We will first explore the aerodynamics and energetics of long-range migration of birds. We will follow this up by examining the similarities between man-made and natural fliers, and the underlying universal scaling that yields the Great Flight Diagrams.

Contents

Migration on Wings

1 Migration on Wings

Animal migration is a fascinating marvel. Millions of years of evolution have conditioned some animals to migrate seasonally for survival of their species. Of all such migrations, the long distance migration of some birds is truly an amazing phenomenon, with some birds traveling extremely long distances nonstop. Bar-tailed godwits (Fig. 1) routinely travel from Alaska and Siberia to Australia and New Zealand and back every year, a total round trip distance of over 22,000 km (13,750 miles), half way around the world. One female godwit, called E7 for a tag on its upper leg, was captured in February 2007 and a tiny radio transmitter was implanted in its abdomen. It was subsequently continuously satellite-tracked by scientists from the U.S. Geological Survey. The godwit left New Zealand on March 17, 2007 and flew nonstop 10,080 km (6,300 miles) to reach China on March 25, 2007. The journey took 8 days and the average flight speed was 14.6 m/s (32.8 mph). The bird stayed in China for 5 weeks, replenishing its fat reserves before it took off on May 1, 2007 across the Sea of Japan and North Pacific, and took 6 days to reach the Yukon-Kusokwim delta in Alaska on May 6, 2007, flying a distance of 7,200 km (4,500 miles) at an average speed of 13.9 m/s (31.3 mph). It summered in Alaska, fattening up on marine worms and clams that it plucked from the mud with its long beak, and breeding. It took off on its fall migration on August 29, 2007, and after a nonstop flight of 11,520 km (7200 miles) lasting 8 days, landed in North Cape, New Zealand on September 7, 2007. The average flight speed was 16.7 m/s (37.5 mph), augmented by strong tail winds up to 11 m/s (25 mph) in the North Pacific. During its flight from Alaska, the bird first headed for Hawaii, then turned southwest toward Fiji before heading south to New Zealand, flying at altitudes as high as 15,000 ft. Scientists estimate the bird to have lost 50% of its body weight, and conjecture that it may have shut off half its brain to nap along the way, as mallard ducks do (Rozell 2007). Its entire flight path is shown in Fig. 2.

Such phenomenal journeys are not confined to this particular bird. Other birds of the species also travel remarkable distances in their annual migration. The USGS team also tracked eight other godwits on what scientists called "extreme endurance

L. Kantha, *Migration on Wings*, SpringerBriefs in Applied Sciences and Technology, 1
DOI: 10.1007/978-3-642-27925-6_1, © The Author(s) 2012

Fig. 1 Bar-tailed godwit
(*Wikipedia*), a wading bird
with a mass less than 1/2 kg.
It flies annually from its
boreal summer habitat in
Alaska to its winter habitat in
New Zealand and back, a
distance of 11,000 km, each
way

flights" of between 6,968 km (4,355 miles) and 11,613 km (7,258 miles) depending
on the route. Figure 3 shows the flight paths of eight bar-tailed godwits during their
fall migration from Siberia to Australia in the October of 2008.

Nor are such migrations confined to godwits. Sooty shearwaters are seabirds
slightly over a kilogram in mass (Fig. 4). They are spectacular long-distance
migrants, traveling north up the western side of the Pacific and Atlantic Oceans at the
end of the nesting season in March–May, reaching sub Arctic waters in June-July,
where they cross from west to east, then returning south down the eastern side of the
oceans in September–October, reaching the breeding colonies in November. They do
not migrate as a flock, but rather as single individuals, Fig. 5 from Proceedings of the
National Academy of Sciences shows the tracks of 19 sooty shearwaters tagged in
early 2005 and tracked for an average of 262 days during their breeding period (light
blue lines) and subsequent migration. Yellow lines show the shearwaters' northward
migration from their breeding sites; orange tracks show the birds' activity in three
northern Pacific foraging zones and their return trip southward.

In the Atlantic Ocean, they cover distances in excess of 14,000 km (8,750 miles)
from their breeding colony on the Falkland Islands (52°S 60°W) north to 60°–70°N in
the North Atlantic Ocean, off north Norway. Distances covered in the Pacific are
similar or larger. Although the Pacific Ocean colonies at 35° to 50°S off New Zealand
are not quite so far south, in moving north to the Aleutian Islands, the longitudinal
width of the ocean makes longer migrations necessary. Recent tagging experiments
have shown that birds breeding in New Zealand may travel 74,000 km in a year,
reaching Japan, Alaska and California, averaging more than 500 km per day.

The Arctic Tern is a small seabird with an average of 110 g in mass and 0.8 m
wing span (Fig. 6). It has a circumpolar distribution, breeding colonially in Arctic
and sub-Arctic regions of Europe, Asia, and North America. The species migrates
from its northern breeding grounds to the oceans around Antarctica and back,
about 38,400 km (24,000 miles) each year. This is one of the longest regular
migrations by any bird. Note however, the journey from one polar region to the
other involves a rest and recharge stop in-between, as seen in Fig. 7. The Arctic
Tern flies as well as glides through the air, performing almost all of its tasks in the

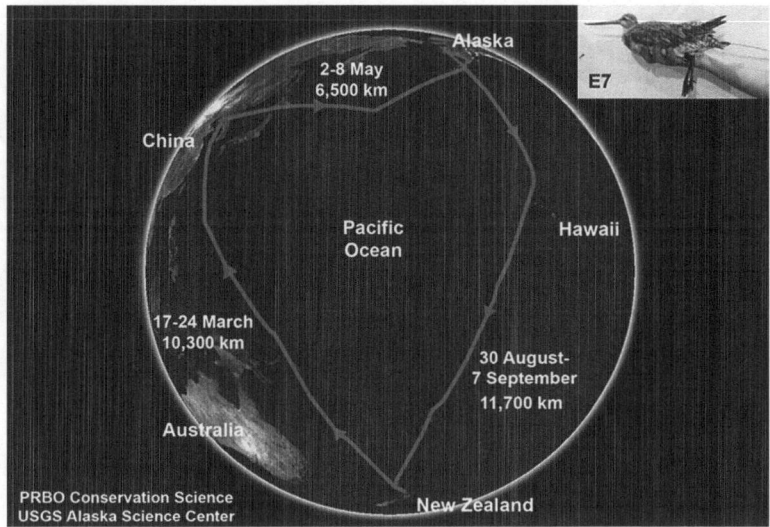

Fig. 2 The epic flight path of a female bar-tailed godwit as it flew across the Pacific in 2007 on its seasonal migration. The bird was tracked by a satellite transmitter implanted in its stomach

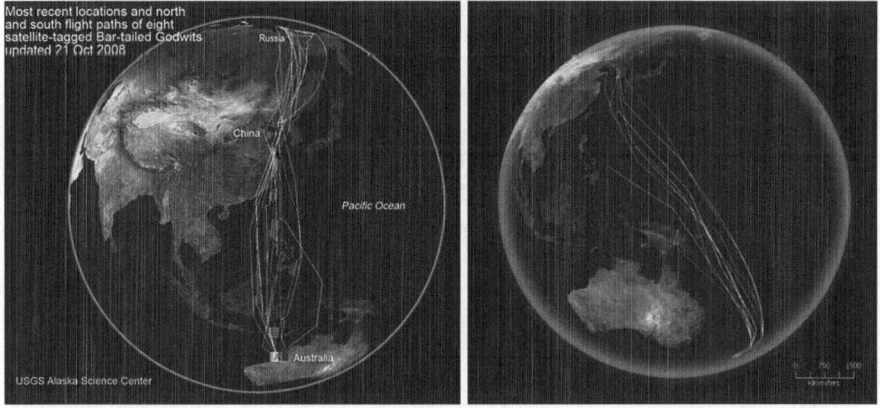

Fig. 3 The flight paths of bar-tailed godwits on their winter migration from Siberian Arctic to Australia in the October of 2008 (*left*) and from New Zealand to Yellow Sea in March 2007, which was 11,026 km (nonstop) and took 9 days

air. It lands once every one to three years (depending on their mating cycle) to nest. Once it has finished nesting, it takes to the sky for another long southern migration. This 19,000 km (about 12,000 miles) journey each way ensures that this bird sees two summers per year and more daylight than any other creature on the planet. The average Arctic Tern in its life will travel a distance equal to going to the moon and back—about 800,000 km (500,000 miles). One example of this bird's remarkable long-distance flying abilities involves an Arctic Tern ringed as an unfledged chick on the Farne Islands, Northumberland, UK, in summer 1982,

Fig. 4 Sooty shearwater
(*Wikipedia*)

which reached Melbourne, Australia, in October 1982, a sea journey of over 22,000 km (14,000 miles) in just 3 months from fledging. Another example is that of a chick ringed in Labrador, Canada, on 23 July 1928. It was found in South Africa four months later (*Wikepedia*). A more recent study by the Arctic Tern Migration Project found tagged Arctic Terns travelling an average of 71,000 km (44,000 miles) in their annual migration between Greenland and Antarctica (Fig. 7), which is 4000 miles more than the Sooty Shearwater and amounts to 2,400,000 km over their 30 year lifetime!

There are numerous other birds that undertake seasonal migration. A tabulation of many of these can be found in Appendices B and C. Because bird migration has been a fascinating subject of interest to both scientists and the public, enormous attention has been lavished on the subject with innumerable scientific articles and popular articles. National Geographic has published articles, along with comprehensive maps of bird migration in the western and eastern hemispheres, which are reproduced in Figs. 8 and 9.

Jacque Perrin's film on "*Winged Migration*" is a real treat for the eyes and is a "must see" for bird migration enthusiasts. Its companion book (Perrin 2003) contains interesting details. A fairly recent book edited by Jonathan Elphick, "*Atlas of Bird Migration*" (Elphick 2009) is an excellent treatment of the subject for the non-scientist types. Hank Tennekes' popular book on "*The Simple Science of Flight*" (Tennekes 2009) has just been updated and is still an easily-read description of bird flight.

In the scientific arena, both ornithologists and aerodynamicists have investigated the phenomenon of bird migration. Rapid advances in technology have enabled radio transmitters to be sufficiently miniaturized and implanted in birds so that they can be tracked continuously by satellites. Examples of this were shown earlier. Thomas Alerstam is an authority on bird migration and his authoritative book "*Bird Migration*" (Alerstam 1990), while a bit dated is still useful to scientists and researchers. Colin Pennycuick is also an authority on the subject and has recently published "*Modelling the Flying Bird*" (Pennycuick 2008), which contains a computer code to model bird migration. The technical literature on bird migration is too voluminous to cite here, but a fairly recent paper by Alerstam and Hedenstrom (1998) and the references therein are a good start. See also Greenewalt (1975), Pennycuick (1969 & 1989) and Shyy et al. (1999).

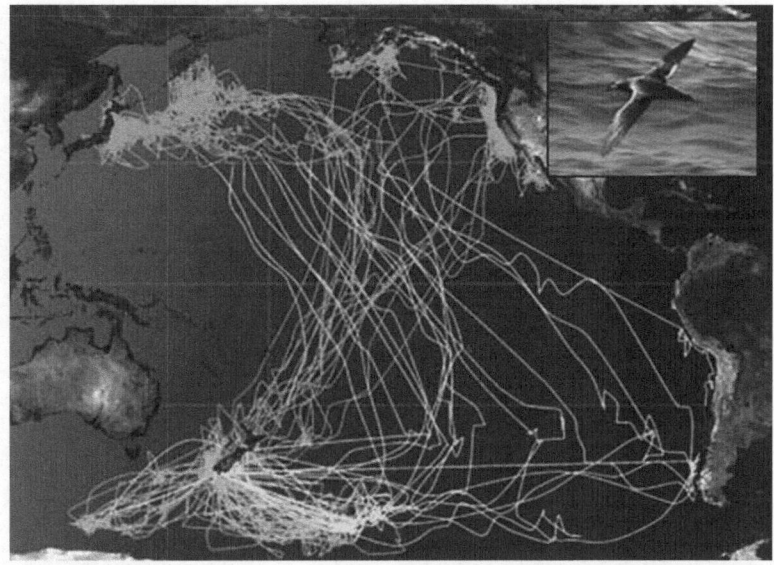

Fig. 5 Flight paths of 19 shearwaters satellite tracked during 2005. *Light blue lines* correspond to the breeding period. *Yellow lines* show the shearwaters' northward migration from their breeding sites; *orange tracks* show the birds' activity in three northern Pacific foraging zones and their return trip southward

Fig. 6 Arctic Tern (*Wikipedia*)

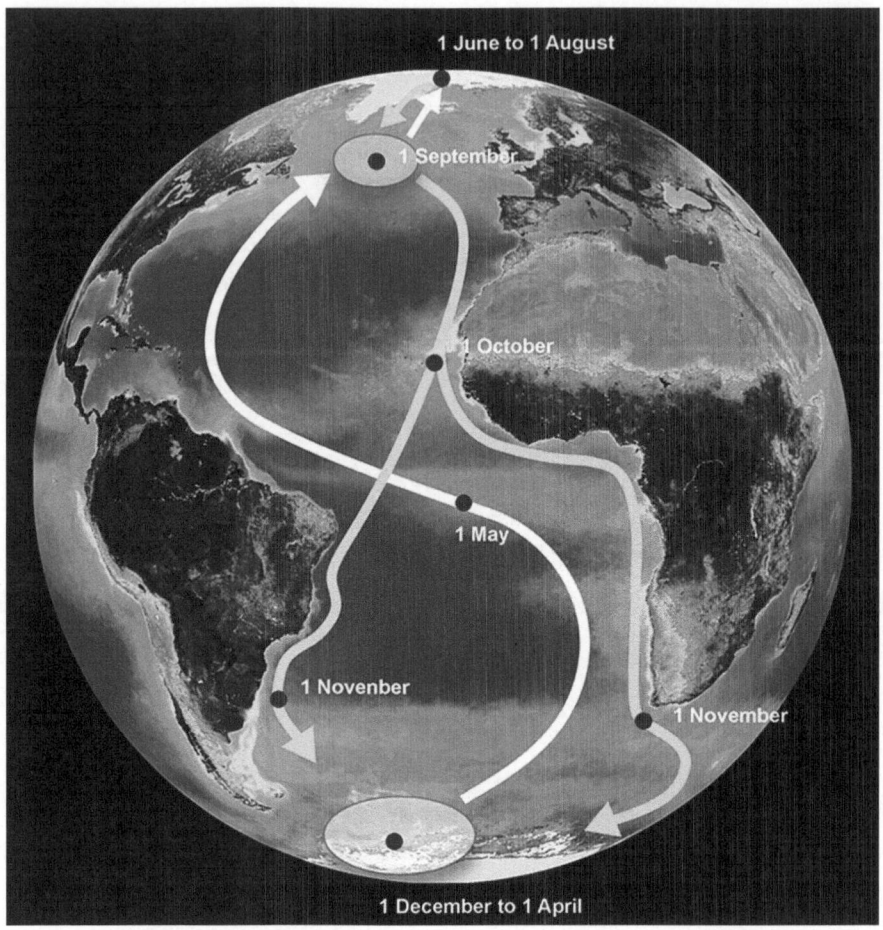

Fig. 7 Arctic Tern Migration Map (courtesy of Greenland Institute of Natural Resources)

Fig. 8 National Geographic wall map of migration in the eastern hemisphere (reproduced with permission)

Fig. 8 (continued)

Fig. 9 National Geographic wall map of bird migration in the Americas (reproduced with permission)

Fig. 9 (continued)

2 Aerodynamics and Energetics of Flight

Ornithologists are interested in physiological and ecological aspects of bird migration, while aerodynamicists have focused on aerodynamics and energetics aspects such as: What are optimum speeds for migration? What sort of fat reserves are needed for carrying out the migration successfully? We will now look at some aspects of winged migration.

First, power requirements for migratory flight. The power required P is given by $P = D \cdot V$, where D is the drag and V is the flight speed. Using the fact that weight W equals the Lift L for level flight, the power per unit body mass can be written as:

$$\frac{P}{m} = \frac{Vg_0}{(L/D)} \tag{1}$$

where L/D is the lift/drag ratio, also known as aerodynamic efficiency. The higher the aerodynamic efficiency, the smaller the power per unit body mass that must be developed by the flight muscles. The higher the flight speed, the higher the power requirement. The L/D ratio depends on the efficiency of lift generation by the wings and the overall aerodynamic drag of the bird. In general, long, narrow wings provide a higher L/D. Such wings have a high aspect ratio, which is defined as $A = b^2/S$, where b is the span and S is the wing area. The drag on the bird consists of two components, the form and skin friction drag, which increases quadratically with flight speed, and the induced drag, which decreases quadratically with flight speed. Induced drag can be thought of as the penalty for lift generation by finite wings, since the induced drag is zero for infinitely long wings ($b \to \infty$). The higher the aspect ratio, the smaller the induced drag. The skin friction drag depends only on the flight Reynolds number, since flight Mach numbers and hence compressibility effects are negligibly small, and the flow very nearly incompressible. The smoother the surface, the lower the skin friction drag. Form drag depends on the shape of the object. The form drag is reduced significantly by minimizing flow separation by streamlining and retracting appendages that might stick out in the flow and cause flow separation. Birds have highly streamlined bodies and wings, and tuck in their feet during flight. Feathers covering their bodies appear to play an important role in reducing the overall drag. Consequently, it is the aspect ratio of their wings that is important in determining their L/D ratio. Song birds have fairly short, wide wings and hence small aspect ratios of around 4, while albatrosses with long, narrow wings can have aspect ratios as high as 25. Ocean birds have aspect ratios typically around 12–14.

The optimum flight speed of a flying object depends on whether the expended power needs to be minimized or the flight distance needs to be maximized. To see this, consider the force balance during level flight.

$$L = \frac{\rho}{2}V^2 S C_L = W$$

$$D = \frac{\rho}{2}V^2 S C_D; \quad P = DV \tag{2}$$

where C_L and C_D are the lift and drag coefficients, which are functions of the wing angle α and the flight Reynolds number $Re = \frac{\rho V b}{\mu}$, where ρ is the air density (which is a function of altitude) and μ is the viscosity. For normal flight speeds, Re is high enough that its influence on skin friction drag does not change much with speed. For an airfoil, the lift coefficient C_L is a function of mainly the angle of attack: $C_L \simeq 0.105\,\alpha$, where α is in degrees. Bird wings are shaped like airfoils and therefore obey this rule. The drag coefficient C_D can be written as:

$$C_D = D_{D0} + C_{Di} = C_{D0} + \frac{C_L^2}{\pi Ae} \tag{3}$$

where subscript 0 denotes the form drag and subscript i, the induced drag. $C_{Di} = \frac{C_L^2}{\pi Ae}$ is the induced drag coefficient, where e is the platform efficiency (also called Oswald efficiency), which depends on the lift distribution on the wing. For the best designed man-made wings, e is typically around 0.85. Since birds flap their wings, which do not generate much lift during the upstroke, the platform efficiency is smaller. A value of 0.75 is reasonable for birds. The drag on the bird and the power required can be written as:

$$D = \frac{\rho}{2}V^2 S \left(C_{D0} + \frac{C_L^2}{\pi Ae}\right) = \left(\frac{\rho S}{2}\right) C_{D0}V^2 + \frac{W^2}{\left(\frac{\rho S}{2}\right)\pi Ae V^2}$$

$$P = \frac{\rho}{2}V^3 S \left(C_{D0} + \frac{C_L^2}{\pi Ae}\right) = \left(\frac{\rho S}{2}\right) C_{D0}V^3 + \frac{W^2}{\left(\frac{\rho S}{2}\right)\pi Ae V} \tag{4}$$

By differentiating with respect to V, it can be shown that minimum drag and minimum power occur for flight speeds given by

$$V_1^2 = V_{minD}^2 = \frac{W}{\left(\frac{\rho S}{2}\right)\sqrt{C_{D0}\pi Ae}} \quad \text{and} \quad V_2^2 = V_{minP}^2 = \frac{W}{\left(\frac{\rho S}{2}\right)\sqrt{3C_{D0}\pi Ae}} \tag{5}$$

respectively. The minimum drag condition also corresponds to minimum energy per unit distance (since $E = DR$, where E is the energy and R is the distance), and hence maximum range: $V_{maxR} = V_{minD}$. So birds must travel at V_{maxR} for making the best use of their fat reserves. On the other hand, they must travel at a slower flight speed V_{minP} for minimum exertion. Note that $V_{maxR} = 1.316\,V_{minP}$.

By differentiating with respect to C_L, it can be shown that minimum drag and minimum power occur for

$$C_{D0} = \frac{C_L^2}{\pi Ae} = C_{Di} \quad \text{and} \quad C_{D0} = \frac{C_L^2}{3\pi Ae} = \frac{1}{3}C_{Di} \tag{6}$$

respectively or equivalently

$$\frac{C_{D0}}{C_{Di}} = 1, \quad \frac{C_{Di}}{C_D} = \frac{1}{2}, \quad C_D = 2C_{D0} \quad \text{and} \quad \frac{C_{D0}}{C_{Di}} = \frac{1}{3}, \quad \frac{C_{Di}}{C_D} = \frac{3}{4}, \quad C_D = 4C_{D0} \tag{7}$$

respectively. The condition for minimum drag D corresponds to maximum possible L/D (since L = W = constant) or maximum aerodynamic efficiency of the flight system, leading naturally to the maximum possible range for the bird. Also

$$\left.\frac{L}{D}\right|_{\min D} = \frac{W}{\left(\frac{\rho S}{2}\right)V_{\min D}^2 (2C_{D0})} \quad \text{and} \quad \left.\frac{L}{D}\right|_{\min P} = \frac{W}{\left(\frac{\rho S}{2}\right)V_{\min P}^2 (4C_{D0})} \tag{8}$$

so that $(L/D)_{\min P} = 0.866 \, (L/D)_{\min D}$.

However, birds have been observed to travel often at flight speeds higher than those corresponding to minimum drag (maximum range). To explain this, it is important to realize that the sustained continuous power output from the flight muscles may permit the bird to travel at a higher speed without tiring it unduly. If we write $C_{Di} = \beta C_{D0}$, β is a decreasing function of flight speed V and the expressions for the flight speed V, drag D, L/D ratio and power P can be written as:

$$V = \beta^{-1/4} \left[\frac{W^{1/2}}{\left(\frac{\rho S}{2}\right)^{1/2}(\pi Ae)^{1/4}}\right] C_{D0}^{-1/4} \tag{9}$$

$$D = (1+\beta)\beta^{-1/2}\left[\frac{W}{(\pi Ae)^{1/2}}\right]C_{D0}^{1/2}; \quad \frac{L}{D} = \frac{\beta^{1/2}}{(1+\beta)}(\pi Ae)^{1/2}C_{D0}^{-1/2} \tag{10}$$

$$P = (1+\beta)\beta^{-3/4}\left[\frac{W^{3/2}}{\left(\frac{\rho S}{2}\right)^{1/2}(\pi Ae)^{3/4}}\right]C_{D0}^{1/4} \quad \text{and}$$

$$\frac{P}{m} = (1+\beta)\beta^{-3/4}\left[\frac{W^{1/2}g_0}{\left(\frac{\rho S}{2}\right)^{1/2}(\pi Ae)^{3/4}}\right]C_{D0}^{1/4} \tag{11}$$

Since $\beta = 3$ for minimum power and 1 for maximum range, Using the above expressions, it is easy to show that

$$P_{\max R} = 1.14 P_{\min P}, \quad V_{\max R} = 1.316 V_{\min P}, \quad \left.\frac{L}{D}\right|_{\min P} = 0.866 \left.\frac{L}{D}\right|_{\max R} \tag{12}$$

More importantly, these expressions allow us to determine the power P and L/D for any given velocity V in terms of the values at either minimum power or maximum range. For example,

$$\frac{L}{D}=\left(\frac{2\beta^{1/2}}{1+\beta}\right)\frac{L}{D}\Bigg|_{\max R} \quad \text{where } \beta = (V_{\max R}/V)^4 \text{ and } P = 0.57\left(\frac{1+\beta}{\beta^{3/4}}\right)P_{\min} \text{ where}$$

$$\beta = (V_{\min P}/V)^4 \tag{13}$$

These expressions can also be used to determine the flight speed for available muscle power. The flight muscles of birds can generate about 100 W/kg of continuous power (twice as much in short bursts) and the mass of the flight muscles is roughly 20–25% of the total body mass of the bird (larger birds have smaller percentage; some birds grow flight muscles before migration). Therefore, the long term sustained power output available for flight from flight muscles is therefore 20–25 W per kg of total body mass. Using

$$\left(\frac{P}{m}\right)_{\text{avilable}} = \frac{(1+\beta)\beta^{-3/4}}{2}\left(\frac{P}{m}\right)_{\max R} \tag{14}$$

it is possible to determine the corresponding value of β and hence the flight speed possible: $V_{\max P} = V_{\max R}\beta^{-1/4}$.

Figure 10 shows the flight speed of various birds that correspond to the sustained muscle power output available for flight plotted against the minimum power (red points), maximum range (blue points) and observed speeds (black points). Both the red and blue points fall above $45°$ straight line, whereas the black points cluster around it. This suggests that migrating birds tend to travel at the sustained power that their flight muscles can produce without undue fatigue, rather than the flight speed that produces maximum range, a condition they are most likely unable to understand and implement, or that corresponding to the minimum effort (power).

For flight at higher altitudes, one may have to account for the lower air density. If the bird does not increase its respiratory rate at higher altitudes, the specific power must be multiplied by the ratio of air density at the flight altitude to that at sea level, which is $\sigma = \dfrac{\rho}{\rho_{SL}}$, since the power output would decrease due to less oxygen (per unit volume) available to burn fat (and other tissues). However, birds could compensate for the lower air density by increased respiration rate and so we will ignore the density change for simplicity.

Figure 11 shows the weight of the bird plotted against the muscle power per unit body mass required for flight at flight speeds corresponding to maximum range as well as the observed flight speeds.

The flight speed of migrating birds (in still air) is of considerable interest. Tennekes (2009) derives a relationship for flight speed as follows: Using Eq. (2), $V = \sqrt{\dfrac{2\left(\frac{W}{S}\right)}{\rho C_L}}$. However, $C_L = 6\alpha$, where α is in radians or equivalently, $C_L = 0.105\alpha$ where α is in degrees. Therefore,

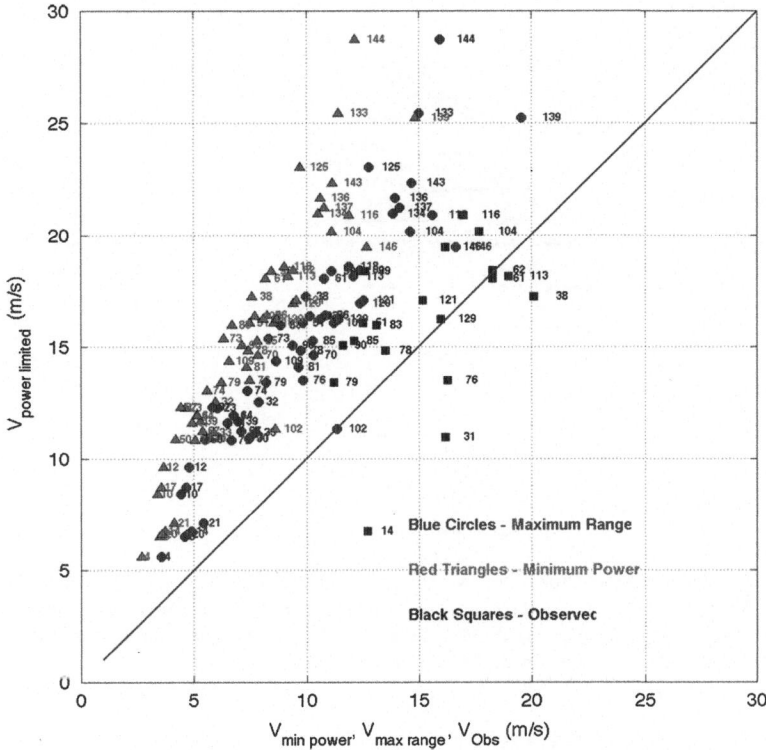

Fig. 10 Fatigue-free muscle power limited flight speed plotted against the flight speeds corresponding to minimum power (*red triangles*), maximum range (*blue circles*) and observed speeds (*black squares*). The data are from Tennekes (2009) listed in Appendix A

$$V = \sqrt{\frac{2\left(\frac{W}{S}\right)}{0.105\,\alpha\,\rho_{SL}\left(\rho/\rho_{SL}\right)}} \tag{15}$$

The best L/D ratio occurs at around 5–6° and the value of 6° is used in Fig. 13, where the bird weight is plotted against the flight speed V. The flight speed depends only on the wing loading W/S. This equation however ignores the influence of L/D and aspect ratio A on flight speed. These parameters can be taken into account as follows:

$$W = qC_{L}S \rightarrow \frac{W}{S} = qC_{L} \rightarrow \left(\frac{W}{S}\right)\left(\frac{1}{C_{D}}\right) = q\left(\frac{C_{L}}{C_{D}}\right) \rightarrow \left(\frac{W}{S}\right)\frac{1}{C_{Di}\left(1 + \frac{C_{D0}}{C_{Di}}\right)}$$

$$= q\left(\frac{L}{D}\right)$$

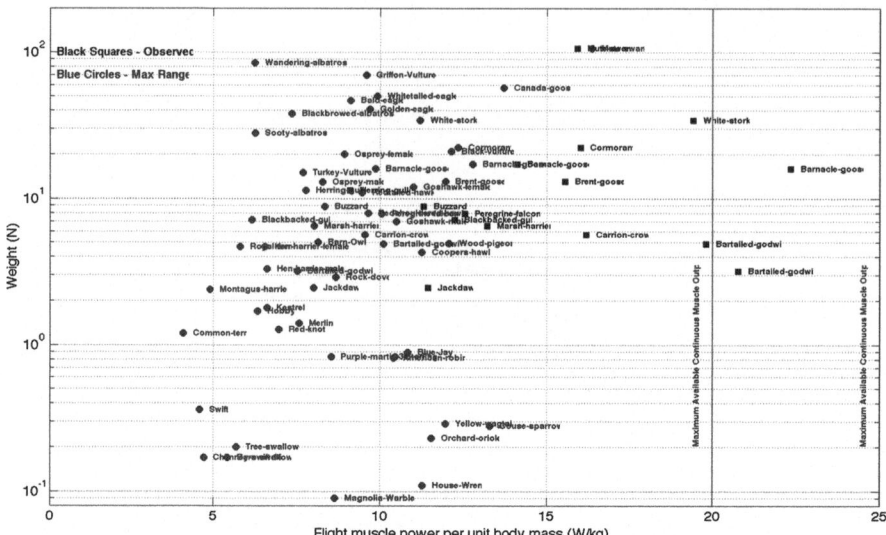

Fig. 11 Power requirement for flight: weight of various birds plotted against the muscle power required for flight per unit body mass at speeds corresponding to maximum range (*blue circles*) as well as observed (*black squares*) flight speeds. The approximate maximum sustained muscle power per unit mass available for flight is shown by the *vertical red lines*

$$\text{Using } D_i = qC_{Di}S = \frac{1}{\pi eq}\left(\frac{W}{b}\right)^2,$$

$$C_{Di} = \frac{1}{\pi eq^2 S}\left(\frac{W}{b}\right)^2 \rightarrow \left(\frac{W}{S}\right)\frac{\pi eq^2 S}{\left(\frac{W}{b}\right)^2\left(1 + \frac{C_{D0}}{C_{Di}}\right)} = q\left(\frac{L}{D}\right)$$

$$\text{Using } AR = \frac{b^2}{S} \rightarrow \frac{\pi eq^2 AR}{\left(\frac{W}{S}\right)\left(1 + \frac{C_{D0}}{C_{Di}}\right)} = q\left(\frac{L}{D}\right) \rightarrow q = \frac{1}{2}\rho V^2 = \frac{\left(\frac{W}{S}\right)\left(\frac{L}{D}\right)\left(1 + \frac{C_{D0}}{C_{Di}}\right)}{\pi e AR}$$

$$\text{Therefore, } V = \sqrt{\frac{2\left(\frac{W}{S}\right)\left(\frac{L}{D}\right)\left(1 + \frac{C_{D0}}{C_{Di}}\right)}{\pi e AR \rho_{SL}\left(\frac{\rho}{\rho_{SL}}\right)}} \qquad (16)$$

$\frac{C_{D0}}{C_{Di}} = 1$ for minimum drag (maximum range), $\frac{1}{3}$ for minimum power. Figure 12 shows also these flight speeds plotted against weight for fliers listed in Appendix A, for whom L/D data are available.

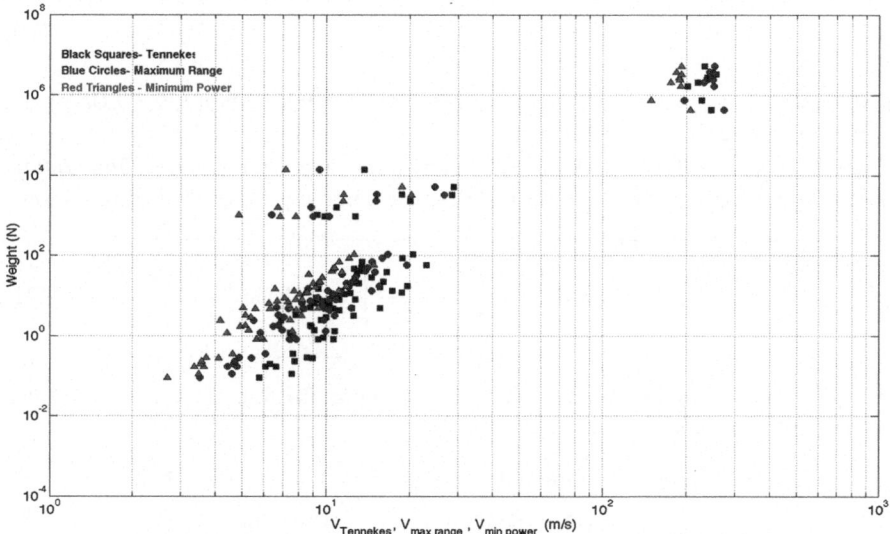

Fig. 12 Weight versus flight speed given by Eq. (15) (*black squares*) and Eq. (16) (*blue circles* for maximum range and *red triangles* for minimum power) for fliers listed in Appendix A, for whom L/D data are available

Note that $D_i = qC_{Di}S = \dfrac{1}{\pi eq}\left(\dfrac{W}{b}\right)^2$ and $\dfrac{D_i}{D_0} = \dfrac{1}{\pi eq^2 C_{D0}AR}\left(\dfrac{W}{S}\right)^2$ (17)

This shows that the induced drag depends only on the span loading whereas the ratio of the induced drag to the form/skin friction drag depends on only the wing loading.

3 Migration Range

The Breguet equation for the range R of migratory birds can be derived as follows:

$$\frac{dR}{dt} = V = \frac{P}{D} = \frac{1}{g_0}\left(\frac{P}{m}\right)\left(\frac{L}{D}\right) = -\frac{1}{g_0}\left(\frac{\dot{m}\,h_c\,\eta}{m}\right)\left(\frac{L}{D}\right) = -\frac{h_c\,\eta}{g_0}\left(\frac{L}{D}\right)\left(\frac{1}{m}\frac{dm}{dt}\right)$$ (18)

so that $dR = -\dfrac{h_c\,\eta}{g_0}\left(\dfrac{L}{D}\right)\dfrac{dm}{m} = -RF\left(\dfrac{dm}{dt}\right)$ (19)

where

$$RF = \frac{h_c \eta}{g_0} \left(\frac{L}{D}\right) \tag{20}$$

is the range factor; h_c is the energy available per unit mass of fat (39.3 MJ/kg) and η the efficiency of conversion of this energy to useful mechanical energy by the flight muscles ($\sim 23\%$ according to Pennyquick and Battly 2003). Integrating, assuming L/D is constant, we get the Breguet range equation of bird migration:

$$R = RF \ln \left(\frac{m_i}{m_f}\right) = (RF) \ln \left(\frac{1}{1 - \frac{m_{fat}}{m_i}}\right) = (RF) \ln \left(1 + \frac{m_{fat}}{m_i}\right) \tag{21}$$

Alternatively, following Tennekes (2009), an approximate equation for range can be derived as follows:

$$\text{Work done} = D \cdot R = \text{Energy produced} = (m_i - m_f) h_c \eta.$$

$$\text{Therefore, } R = \frac{(m_i - m_f) h_c \eta}{D} \left(\frac{L}{W}\right) = \frac{(m_i - m_f)}{m g_0} \left(\frac{L}{D}\right) h_c \eta \tag{22}$$

$$\text{Using } m = m_{av}, R = \frac{2 h_c \eta}{g_0} \left(\frac{L}{D}\right) \frac{(m_i - m_f)}{(m_i + m_f)} = 2(RF) \frac{\left(\frac{m_i}{m_f} - 1\right)}{\left(\frac{m_i}{m_f} + 1\right)}.$$

$$\text{Therefore, } \quad R = (RF) \frac{\left(\frac{m_{fat}}{m_i}\right)}{\left[1 - 0.5 \left(\frac{m_{fat}}{m_i}\right)\right]} = RF \left(\frac{\frac{m_{fat}}{m_f}}{1 + \frac{m_{fat}}{2 m_f}}\right) \tag{23}$$

This equation underestimates R slightly for large values of m_{fat}/m_f. Note that the usual limit on m_{fat}/m_f is around 1.0, although after exhausting the fat reserves under adverse flight conditions, the bird may have to use up muscle mass for power generation, increasing the extreme limit to around 1.2.

Both approaches assume L/D = constant during the migration flight. This is not correct for birds that undergo significant mass changes during their migration, i.e., for long distance nonstop migrants such as the bar-tailed godwit. A range equation for this situation can be readily derived by recognizing $RF = RF_i + \left(\frac{m_i - m}{m_i - m_f}\right)$ $(RF_f - RF_i)$. Then Eq. (19) becomes

$$dR = -\left[RF_i + \left(\frac{m_i}{m_i - m_f}\right)(RF_f - RF_i)\right]\left(\frac{dM}{m}\right) - \left(\frac{RF_f - RF_i}{m_i - m_f}\right) dm$$

Integrating this results in:

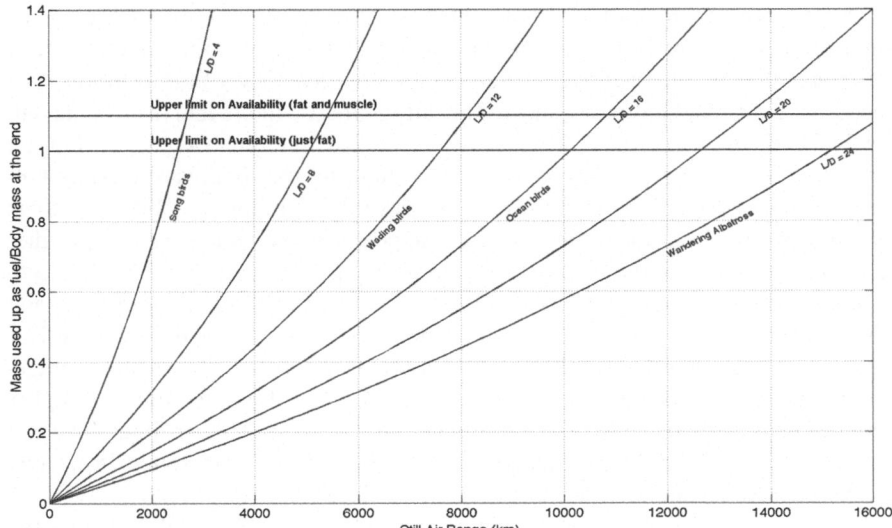

Fig. 13 The fuel reserves needed for a given still air range as a function of L/D. The *lower red line* corresponds to using up just the fat reserves, whereas the *upper* one to using up some of the flight muscles as well

$$
\begin{aligned}
R &= \left[RF_i + \left(\frac{m_i}{m_i - m_f} \right)(RF_f - RF_i) \right] \ln\left(\frac{m_i}{m_f} \right) - (RF_f - RF_i) \\
&= \left[RF_i + \left(\frac{m_i}{m_{fat}} \right)(RF_f - RF_i) \right] \ln\left(\frac{1}{1 - \dfrac{m_{fat}}{m_i}} \right) - (RF_f - RF_i) \\
&= \left[RF_i + \left(1 + \frac{m_f}{m_{fat}} \right)(RF_f - RF_i) \right] \ln\left(1 + \frac{m_{fat}}{m_f} \right) - (RF_f - RF_i)
\end{aligned}
\tag{24}
$$

Take the case of a bar-tailed godwit mentioned in the introduction on its nonstop migration from Alaska to New Zealand, for which the observed value of $m_{fat}/m_i = 1$ and therefore $R = 0.386\ RF_f + 0.307\ RF_i$. Let us suppose that initially it is so fat that $RF_i = 0.8\ RF_f$. Then $R = 0.617\ RF_f$ instead of $R = 0.693\ RF_f$ if it had the same L/D ratio throughout the flight. This is a 11% reduction. Using Eq. (23) instead and an average value for $RF = 0.5(RF_i + RF_f)$, we would have gotten $R = 0.667\ RF_f$ for constant L/D and $R = 0.6\ RF_f$ for varying L/D. The error in using the approximate formula is about 3.8 and 2.8%, respectively, which is quite small in view of the uncertainties involved in estimating the range factor.

Figure 13 shows the fuel reserves required for a given migration distance for various values of L/D, assuming migration occurs at altitudes of less than 500 m and conversion efficiency η is 23%. Clearly, the upper limit on nonstop migration distance is around 2,000 km for song birds, whereas ocean birds can travel around 10,000 km nonstop.

Irrespective of which of the above three equations are used for computing the range of the migrating bird, it is clear that the range depends on four factors: 1. The aerodynamic efficiency L/D, 2. The efficiency of conversion of fat to mechanical energy η, 3. The flight altitude, and 4. The ratio of the available fuel reserve mass to the final body mass m_{fat}/m_f. Taking the bar-tailed godwit as an example, the longest distance this bird travels nonstop is 11,500 km during the southward migration to New Zealand and 7,500 km during its northward migration to China. Godwits migrating from Siberia to northern Australia appear to make a refueling stop in either Korea or China so that the maximum distance they travel nonstop is from Korea to northern Australia around 6,700 km. Since the journey from Alaska to New Zealand appears to get some assist from strong tail winds until roughly 30° latitude, the still air range must be greater than 7,500 km and less than 11,500 km. Taking the average, the still air range must be around 9,500 km. Accounting for some travel at higher altitudes (this particular bird appears to have reached 15,000 ft at times), we can round this off to 10,000 km. This requires an average L/D ratio of around 15, a value this bird appears to achieve, according to the satellite tracking.

The breeding habitat in Yukon delta is at roughly 60°N, 162.5°W, whereas the winter habitat in New Zealand is at roughly 35°S, 173°E. Siberian summer habitat is around 70–75°N, 140–160°E, whereas the Australian wintering habitat is at 15°S, 125°E (Yalu river delta in Korea is around 40°N, 124°E and Yangtze delta in China is 30.5°N, 121°E). Incidentally, the great circle distance between two points whose longitudes and latitudes are λ_1, λ_2 and ϕ_1, ϕ_2 (in radians) is

$$D_c = R_E \cos^{-1}[\cos \phi_1 \cos \phi_2 \cos(\lambda_1 - \lambda_2) + \sin \phi_1 \sin \phi_2] \qquad (25)$$

in kilometers, where $R_E = 6{,}371$ km is the radius of Earth.

Pennycuick and Battley (2003) did a thorough study of great knots migrating from northwest Australia (18°S, 122.4°E) to Chongming Island at the mouth of Yangtse river in China (30.8°N, 121.5°E) flying 5,420 km nonstop in about 4 days in early April 1998. They collected physiological data on 10 birds before and after migration. The wing span $b = 0.586$ m, the wing area $S = 0.0397$ m^2 and therefore the aspect ratio $A = 8.65$ on the average. The initial mass of the birds was 233.4 g, out of which 89.8 g were fat. The final mass was 125.0 g, but not all fat was consumed during the flight. About 10.7 g of fat remained. This means that the birds consumed 30.3 g of muscle and other tissue! Pennycuick and Battley (2003) used a complex code to compute the likely range of the birds and showed that the birds had the capability to fly 7,040 km, if they used up all their fat reserves, discounting the notion that tail winds are essential to such long distance migration. These results are of course dependent on various assumptions made in the simulation, the most critical of which is the body drag coefficient, for which they assumed a value of 0.1 instead of values twice as high indicated by wind tunnel experiments on frozen birds. They argued that a lower value is more realistic for live birds.

The uncertainty associated with the body drag in the above case study is essentially equivalent to the uncertainty in the L/D ratio of the bird. Consequently, the use of a simpler method based on Eq. (21), (23) or (24) is well justified.

However, it is necessary to account for burning of the body tissue, since the value of h_c for protein is 18.3 MJ/kg less than half of the 39 MJ/kg value for fat. Equation (18) must be modified to:

$$\frac{dR}{dt} = -\frac{1}{g_0}\left(h_c^{fat}\dot{m}_{fat} + h_c^{prot}\dot{m}_{prot}\right)\left(\frac{\eta}{m}\right)\left(\frac{L}{D}\right)$$

which is difficult to integrate analytically unless it is assumed that the tissues are consumed after the fat, which may not be strictly speaking correct (however, the error is probably within other uncertainties in this type of calculation, e.g., L/D). If m_{fat} is the mass of fat and m_{tis} is the mass of body muscle and other tissues consumed during the journey,

$$R = RF_f \ln\left(\frac{m_i}{m_i - m_{fat}}\right) + RF_t \ln\left(\frac{m_i - m_{fat}}{m_i - m_{fat} - m_{tis}}\right) \tag{26}$$

Equivalently,

$$R = RF_f \ln\left(\frac{1 + \dfrac{m_{fat}}{m_f} + \dfrac{m_{tis}}{m_f}}{1 + \dfrac{m_{tis}}{m_f}}\right) + RF_t \ln\left(1 + \frac{m_{tis}}{m_f}\right) \tag{27}$$

where $$RF_f = \frac{h_c^{fat}\eta}{g_0}\left(\frac{L}{D}\right), \quad RF_t = \frac{h_c^{tis}\eta}{g_0}\left(\frac{L}{D}\right) \tag{28}$$

This equation is useful in computing the extra range attainable by burning body tissue after the bird has exhausted the fat reserve. Applying this equation to great knots, assuming an average value of 10 for L/D, the range is 4,440 + 1,020 = 5,460 km, close to the observed value.

Is L/D = 10 the correct average value for great knots? The above calculation assumed sequential burning of fat and tissue. Analytical Breguet range equation cannot be derived for simultaneous burning. However Eq. (23) may be modified to account for both cases:

$$R = \frac{2}{g_0}\left(\frac{L}{D}\right)\frac{\left(m_{fat}h_c^{fat} + m_{tis}h_c^{tis}\right)\eta}{(m_i + m_f)}$$

so that

$$R = \frac{RF_f\left(\dfrac{m_{fat}}{m_i}\right) + RF_t\left(\dfrac{m_{tis}}{m_i}\right)}{\left[1 - 0.5\left(\dfrac{m_{fat}}{m_i}\right) - 0.5\left(\dfrac{m_{tis}}{m_i}\right)\right]} = \frac{RF_f\left(\dfrac{m_{fat}}{m_f}\right) + RF_t\left(\dfrac{m_{tis}}{m_f}\right)}{\left(1 + \dfrac{m_{fat}}{2m_f} + \dfrac{m_{tis}}{2m_f}\right)} \tag{29}$$

Using Eq. (29) results in required average L/D of 11.4 for the observed 5,420 km range. Alternatively, if the weighted value of RF: $RF_W = \left(\dfrac{RF_f m_{fat} + RF_t m_{tis}}{m_{fat} + m_{tis}} \right)$ is used in Eq. (21)

$$R = RF_W \ln\left(\frac{m_i}{m_f}\right) = RF_W \ln\left(\frac{1}{1 - \frac{m_{fat}}{m_i} - \frac{m_{tis}}{m_i}}\right) = RF_W \ln\left(1 + \frac{m_{fat}}{m_f} + \frac{m_{fat}}{m_f}\right) \qquad (30)$$

which leads to a required L/D of 11.1; Either value is quite plausible for a great knot.

Some "What if" questions can be posed. How much farther could the great knot travel, if it used up the remaining fat reserve? This is simply $RF_f \ln\left(\dfrac{m_f}{m_f - m_{fat}^{remaining}}\right) = 911$ km for a total of 6,331 km. What if all the stored fat was used up first and then the tissue? Using Eq. (27), R = 6,371 roughly similar.

Pennycuick and Battly (2003) also present data for the bar-tailed godwit. The birds starting from Alaska (wing span b = 0.748 m, the wing area S = 0.0568 m^2 and aspect ratio A = 9.85 on the average) had an initial mass of 367 g, out of which 201 g were fat. Assuming that all the fat is consumed during the winter flight to New Zealand, Eq. (21) gives a range of 10,880 km for an average L/D ratio of 15, quite adequate for the migration. If additional range were needed under adverse flight conditions, burning some 20 g of tissue would extend the range by about 830 km. Thus, it appears that some tail winds may not be needed for the godwit to undertake this migration. Of course, birds do have the extraordinary ability to select favorable wind and weather conditions for departure as well as flight altitudes with favorable tail winds (Liechti 2006), but they are in no way capable of predicting what wind conditions they are likely to encounter over a flight lasting many days along a route thousands of kilometers long. Consequently, if persistently adverse flight conditions are encountered en route, successful completion of migration may not be possible!

As Pennycuick and Battly (2003) point out, it is the fat fraction, the ratio of the fat mass to the mass of the bird that is of critical importance to the range. Godwits starting on their migration from Alaska to New Zealand use their remarkable ability to make sure this ratio is the highest possible, by actually shrinking the tissues such as their digestive system not needed for the flight. Pennycuick and Battly (2003) found that the northward bound birds did not do so and while their fat reserves were roughly the same (191 g fat, 446 g total mass), the ratio was 0.43 compared to 0.55 for the southward bound birds. This is probably because the northward bound birds tend to make a refueling stop in northern Australia or China, the latter at a distance of only 7,200 km, which is well within the 7,670 km range achievable with the smaller fat fraction.

It is clear that for a given fat to body mass ratio, the major parameter that governs the migration range of birds is their L/D ratio, which while being a function of the wing aspect ratio for the most part, is also a function of the aerodynamic drag characteristics of the bird itself. Unfortunately, this is a parameter that is not available for most birds. Generally speaking, larger wing spans correspond to higher L/D ratios. Figure 14 shows L/D of birds plotted

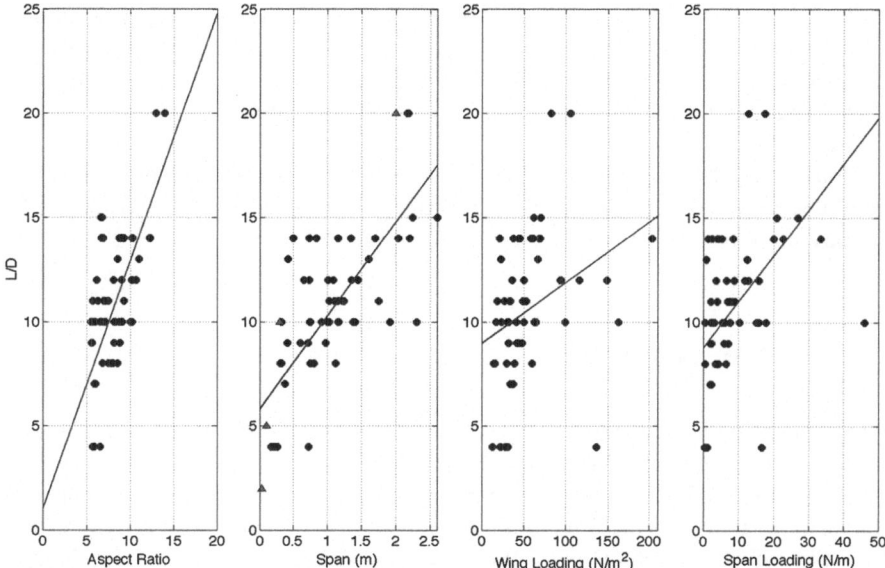

Fig. 14 L/D ratio of various birds plotted against their aspect ratio, wing span, wing loading and span loading from data from Tennekes (2009). In the second panel are also shown data (*red triangles*) inferred from Templin (2000). *Blue lines* show robust least square fits to data

against their aspect ratio, wing span, wing loading and span loading, from data in Tennekes (2009) listed in Appendix A. Also shown are data points (red triangles) inferred from Figs. 1 and 9 of Templin (2000), who suggests that the wing span is the principal determinant of the migration range of birds. Clearly, of all the four quantities, L/D is best correlated with aspect ratio. However, the correlation is imperfect, with significant scatter and hence this result must be used with caution.

L/D ratio is beter correlated with the wetted aspect ratio ($A_{wet} = b^2/S_{wet}$). Decades of aircraft design experience attest to this. For example, AVRO Vulcan bomber had an aspect ratio of just 3, small compared to that of Boeing B-47, which had an aspect ratio of 9.4. Yet, both had L/D ratio of around 17. This is because Vulcan was a flying wing with a wetted area of 892 m^2 and the wing area 320 m^2, compared to B47, which was a conventional design with a wetted area of 1,050 m^2 and the wing area 132 m^2. The ratio of wetted area to wing area was therefore 2.8 for Vulcan and 7.9 for B47, yielding wetted aspect ratio A_{wet} of 1.1 and 1.2 for Vulcan and B47, respectively. Consequently, the L/D ratios were roughly the same for both (Raymer 2006). Note that a flying wing would have a wetted area to wing area ratio of slightly more than 2.

Unfortunately, wetted areas are not easily available for birds and bats (let alone insects). Obtaining the ratio of wetted surface area to the wing area along with a better and more comprehensive data on L/D of migrating birds is therefore, highly desirable and should be a priority for future research. Taking two extreme examples, the albatross and the house wren, the albatross has an aspect ratio of 13 and a L/D ratio of 20–24, while house wren has an aspect ratio of about 5 and L/D ratio of about 4.0. The birds appear to have a wetted area to wing area ratio of approximately 2.3–2.5. The wetted aspect ratios are 5.6 and 2.0, respectively, for albatross and wren. Using these two extreme limits and assuming a linear relationship between the L/D ratio and the wetted aspect ratio A_{wet}, we get

$L/D = 5A_{wet} - 6$

Until more data become available, the migration range capability of a bird should be regarded as an approximate function of its wetted aspect ratio and the fat to body mass ratio at the beginning of migration.

4 Formation Flight

Many migrating birds tend to travel together in a V, J or W shaped formation, or echelon formation of anywhere from a few birds to over a 100. Canada geese, snow geese, brants, sandhill cranes, whooping cranes and american white pelicans are among the birds that fly in formation. The question is why do they do that? There have been two schools of thought on this. One, consisting of mainly ornithologists, advances the proposition that these formations are simply to promote better visual contact and communication with each other without collisions among the flock. The other, consisting of mainly aerodynamicists, holds the view that formation flights are energetically profitable, since they lead to significant reductions in drag, and hence the power and energy expended by each bird in the formation compared to its solo flight. It is likely that both are equally good reasons for the birds to engage in formation flight. The task is then to compute the advantages resulting from formation flight of a given shape. To do so, one needs to appeal to Prandtl's famous lifting line theory. In this theory, each wing can be replaced by a horseshoe vortex with the circulation around the vortex dependant on the lift on the wing.

In a formation flight, every wing flies in the upwash induced by every other wing of the formation. The total upwash w_m induced on a given wing M is the sum of the individual upwashes w_{mn} induced by all other wings in the formation and in general, varies along the span y of that wing. The average value of the upwash is $\bar{W}_m = \sum_{n=1}^{N_w} \bar{W}_{mn}$ where N_w is the total number of wings in the formation. Because of this upwash, the direction of lift us turned forward through an angle $\Delta\alpha = \dfrac{\bar{w}_m}{V}$, which causes the drag to be reduced by $\Delta D_m = L_m\left(\dfrac{\bar{w}_m}{V}\right)$. The corresponding reduction in the power required for flight is $\Delta P_m = V\Delta D_m = L_m\bar{w}_m$, where L_m is the lift and V is the flight speed. The fractional reduction in flight power is

$$\eta_m = \frac{\Delta P_m}{P_m} = \left(\frac{L}{D}\right)_m \frac{\bar{w}_m}{V} \tag{31}$$

Consider wing M. Its semi-span is b_m and its center is located at (x_m, y_m). x is the coordinate in the flight direction and y is the coordinate in the spanwise direction. Let N be the inducing wing with semi-span b_n with its center located at (x_n, y_n). Let b be the normalizing length, the reference span. All the distances are normalized by refernce span b. The normalized distance between the wing centers in the flight direction is $X_{mn} = \frac{X_m - X_n}{b}$. The normalized distance between the wing centers in the spanwise direction is $Y_{mn} = \frac{Y_m - Y_n}{b}$. The distance between wing tips is $\Delta y_{mn} = Y_{mn} - \frac{b_m + b_n}{2}$. The normalized distance between the wing centers in the vertical direction is $Z_{mn} = \frac{Z_m - Z_n}{b}$. The mean upwash $\bar{w}$ induced on wing M by wing N is given by $\frac{\bar{w}_{mn}}{V} = \frac{C_{Ln}}{\pi A_n} \bar{f}_{mn}$, where V is the flight speed, C_L is the lift coefficient and A is the aspect ratio. The parameter $\bar{f}_{mn}$ is a pure geometric function dependent on the wing sizes and relative location and following Higdon and Corrsin (1978) can be determined as follows:

Consider the effect of wing N on wing M. Assume wing N can be replaced by a horse-shoe vortex consisting of two trailing vortices separated by a distance $2a_n$ and a bound vortex. Note that $a_n = \frac{\pi}{4} b_n$. Choosing to place the origin at the midspan of wing N, the vertical velocity induced at point P (x, y, z) can be written as:

$$w(x,y,z) = \frac{\Gamma_n}{2\pi} \left\{ \left[\frac{y - a_n}{(y - a_n)^2 + z^2} \left(1 + \frac{x}{\ell_1} \right) \right] - \left[\frac{y + a_n}{(y + a_n)^2 + z^2} \left(1 + \frac{x}{\ell_2} \right) \right] \right.$$
$$\left. + \left[\frac{x(y - a_n)}{\ell_1(x^2 + z^2)} - \frac{x(y + a_n)}{\ell_2(x^2 + z^2)} \right] \right\}$$

for $|y| > a_n$

$$w(x,y,z) = \frac{\Gamma_n}{2\pi} \left\{ \left[-\frac{a_n - y}{(a_n - y)^2 + z^2} \left(1 + \frac{x}{\ell_1} \right) \right] - \left[\frac{y + a_n}{(y + a_n)^2 + z^2} \left(1 + \frac{x}{\ell_2} \right) \right] \right.$$
$$\left. + \left[\frac{x(y - a_n)}{\ell_1(x^2 + z^2)} - \frac{x(y + a_n)}{\ell_2(x^2 + z^2)} \right] \right\} \quad \text{for } |y| < a_n$$

where the first term in the curly brackets is due to the left vortex L, the second term due to the right vortex R and the third term is due to the bound vortex and $\ell_1 = \left[x^2 + z^2 + (y - a_n)^2 \right]^{1/2}$; $\ell_2 = \left[x^2 + z^2 + (y + a_n)^2 \right]^{1/2}$. Positive w denotes upwash and the negative value, downwash. Γ_n is the circulation of the horse-shoe vortex. Assuming the wings have elliptic lift distribution, $\Gamma_n = \Gamma_n^{max}$ since the span is taken as $2a_n$ and not the actual value $2b_n$. Wing M is located at a distance x in the streamwise direction, z in the vertical direction and y_{mn} in the spanwise direction. The induced drag due to the vertical velocity w along the span of the wing M due to wing N is (King and Gopalarathnam 2005):

$$D_{mn} = -\frac{\rho}{2}\Gamma_m \int\limits_{y_{mn}-b_m}^{y_{mn}+b_m} w(x,y,z)dy. \tag{33}$$

The upper and lower limits correspond to the actual physical span of wing M and therefore, the circulation Γ_m must correspond to the average value $\Gamma_m = \Gamma_m^{av}$ and not the maximum value $\Gamma_m = \Gamma_m^{max}$. Substituting Eq. (32) in Eq. (33):

$$D_{mn} = \frac{\rho\Gamma_m\Gamma_n}{4\pi}\left\{\frac{1}{2}\ln\left[\frac{(y+a_n)^2+z^2}{(y-a_n)^2+z^2}\right]+\frac{1}{2}\ln\left[\frac{(\ell_2-x)(\ell_1+x)}{(\ell_2+x)(\ell_1-x)}\right]+\left[\frac{x(\ell_2-l_1)}{(x^2+z^2)}\right]\right\}\Bigg|_{y=y_{mn}-b_m}^{y=y_{mn}+b_m} \tag{34}$$

for both $|y| > a_n$ and $|y| < a_n$. The first term is the induced drag if the trailing vortices of wing N were to extend from the spanwise position of wing M (which they don't), the second term is due to the missing pieces of the vortices of wing N that would have made them extend to semi-infinity from the spanwise position of the wing M, and the last term is due to the bound vortex of wing N. Note that for $x = x_{mn} = 0$, both the second and third terms vanish and only the contribution due to the semi-infinite trailing vortices of wing N remain. This comes in handy when applying Munk's stagger theorem, which states that in a formation, lifting surfaces can be translated in the streamwise direction arbitrarily, without changing the TOTAL induced drag of the formation (although individual values depend on the exact location of individual wings in the space). This means all the wings in the formation can be moved to a single vertical plane so that $x_{mn} = 0$ and the simplified form of Eq. (34) can then be used to calculate the total induced drag of the formation. Substituting the upper and lower limits:

$$D_{mn} = \frac{\rho\Gamma_m\Gamma_n}{4\pi}f_{mn}$$

$$f_{mn} = \frac{1}{2}\ln\left\{\frac{z^2+[y_{mn}+(a_n+b_m)]^2}{z^2+[y_{mn}-(a_n-b_m)]^2}\cdot\frac{z^2+[y_{mn}-(a_n+b_m)]^2}{z^2+[y_{mn}+(a_n-b_m)]^2}\right\}$$

$$+\frac{1}{2}\ln\left\{\frac{(\ell_2^--x)(\ell_1^++x)}{(\ell_2^++x)(\ell_1^+-x)}\cdot\frac{(\ell_2^++x)(\ell_1^--x)}{(\ell_2^--x)(\ell_1^-+x)}\right\}$$

$$+\left\{\frac{x}{(x^2+z^2)}[(\ell_2^+-\ell_2^-)-(\ell_1^+-\ell_1^-)]\right\} \tag{35}$$

$$\text{where}\quad \ell_2^+ = \left\{x^2+z^2+[y_{mn}+(a_n+b_m)]^2\right\}^{1/2};$$

$$\ell_2^- = \left\{x^2+z^2+[y_{mn}+(a_n-b_m)]^2\right\}^{1/2};$$

$$\ell_1^+ = \left\{x^2+z^2+[y_{mn}-(a_n-b_m)]^2\right\}^{1/2}; \tag{36}$$

$$\ell_1^- = \left\{x^2+z^2+[y_{mn}-(a_n-b_m)]^2\right\}^{1/2}$$

Once again, the first curly brackets are due to semi-infinite vortices, the second due to missing pieces and the third due to the bound vortex. Both the second and third terms vanish for x = 0. All the quantities in curly brackets involve distances and hence can be normalized by the reference span b so that

$$\bar{f}_{mn} = \frac{1}{2}\ln\left\{\frac{z_{mn}^2 + [Y_{mn} + (a_n + b_m)]^2}{z_{mn}^2 + [Y_{mn} - (a_n - b_m)]^2} \cdot \frac{z_{mn}^2 + [Y_{mn} - (a_n + b_m)]^2}{z_{mn}^2 + [Y_{mn} + (a_n - b_m)]^2}\right\}$$

$$+ \frac{1}{2}\ln\left\{\frac{(\ell_2^+ - X_{mn})(\ell_1^+ + X_{mn})}{(\ell_2^+ + X_{mn})(\ell_1^+ - X_{mn})} \cdot \frac{(\ell_2^- + X_{mn})(\ell_1^- - X_{mn})}{(\ell_2^- - X_{mn})(\ell_1^- + x_{mn})}\right\}$$

$$+ \left\{\frac{X_{mn}}{(x_{mn}^2 + z_{mn}^2)}\left[(\ell_2^+ - \ell_2^-) - (\ell_1^+ - \ell1_1^-)\right]\right\}$$

$$\ell_2^+ = \left\{X_{mn}^2 + z_{mn}^2 + [Y_{mn} + (a_n + b_m)]^2\right\}^{1/2};$$

$$\ell_2^- = \left\{X_{mn}^2 + z_{mn}^2 + [Y_{mn} + (a_n - b_m)]^2\right\}^{1/2}$$

and $\qquad\qquad\qquad\qquad\qquad\qquad\qquad\qquad\qquad$ (37)

$$\ell_1^+ = \left\{X_{mn}^2 + z_{mn}^2 + [Y_{mn} - (a_n - b_m)]^2\right\}^{1/2};$$

$$\ell_1^- = \left\{X_{mn}^2 + z_{mn}^2 + [Y_{mn} - (a_n - b_m)]^2\right\}^{1/2}$$

where quantities a_n, b_m, ℓ_2^+, ℓ_2^-, ℓ_1^+, ℓ_1^- are now dimensionless having been normalized by reference span b. Note that

$$D_{mn} = \frac{\rho\Gamma_m\Gamma_n}{4\pi}\bar{f}_{mn} \qquad\qquad\qquad (38)$$

In the case where there is no vertical separation between wings, $Z_{mn} = 0$, and we get

$$\bar{f}_{mn} = \ln\left\{\frac{[Y_{mn} + (a_n + b_m)]}{[Y_{mn} - (a_n - b_m)]} \cdot \frac{[Y_{mn} - (a_n + b_m)]}{[Y_{mn} + (a_n - b_m)]}\right\}$$

$$+ \frac{1}{2}\ln\left\{\frac{(\ell_2^+ - X_{mn})(\ell_1^+ + X_{mn})}{(\ell_2^+ + X_{mn})(\ell_1^+ - X_{mn})} \cdot \frac{(\ell_2^- + X_{mn})(\ell_1^- - X_{mn})}{(\ell_2^- - X_{mn})(\ell_1^- + X_{mn})}\right\}$$

$$+ \left\{\frac{1}{(X_{mn})}\left[(\ell_2^+ - \ell_2^-) - (\ell_1^+ - \ell_1^-)\right]\right\}$$

$$\ell_2^+ = \left\{ X_{mn}^2 + [Y_{mn} + (a_n + b_m)]^2 \right\}^{1/2}; \quad \ell_2^- = \left\{ X_{mn}^2 + [Y_{mn} + (a_n - b_m)]^2 \right\}^{1/2}$$

with

$$\ell_1^+ = \left\{ X_{mn}^2 + [Y_{mn} - (a_n - b_m)]^2 \right\}^{1/2}; \quad \ell_1^- = \left\{ X_{mn}^2 + [Y_{mn} - (a_n - b_m)]^2 \right\}^{1/2}$$

(39)

Note that $\bar{f}_{mn} \neq 0$ for m = n. These expressions are similar to that presented by Hummel (1983) except for a multiplying factor. The terms due to missing trailing vortex pieces however do not agree:

$$\frac{\pi^2}{2} \left(\frac{B_m}{B_n} \right) \bar{f}_{mn} = \ln \left[\frac{(Y_{mn} - a_2)(Y_{mn} + a_2)}{(Y_{mn} - a_1)(Y_{mn} + a_1)} \right]$$

$$+ \ln \left[\frac{Y_{mn} + a_2 - X_{mn} + \sqrt{(Y_{mn} + a_2)^2 + X_{mn}^2}}{Y_{mn} + a_2 + X_{mn} + \sqrt{(Y_{mn} + a_2)^2 + X_{mn}^2}} \right]$$

$$+ \ln \left[\frac{Y_{mn} - a_2 - X_{mn} + \sqrt{(Y_{mn} - a_2)^2 + X_{mn}^2}}{Y_{mn} - a_2 + X_{mn} + \sqrt{(Y_{mn} - a_2)^2 + X_{mn}^2}} \right]$$

$$- \ln \left[\frac{Y_{mn} + a_1 - X_{mn} + \sqrt{(Y_{mn} + a_1)^2 + X_{mn}^2}}{Y_{mn} + a_1 + X_{mn} + \sqrt{(Y_{mn} + a_1)^2 + X_{mn}^2}} \right]$$

$$- \ln \left[\frac{Y_{mn} - a_1 - X_{mn} + \sqrt{(Y_{mn} - a_1)^2 + X_{mn}^2}}{Y_{mn} - a_1 + X_{mn} + \sqrt{(Y_{mn} - a_1)^2 + X_{mn}^2}} \right]$$

$$+ \frac{1}{X_{mn}} \left[\sqrt{(Y_{mn} + a_2)^2 + X_{mn}^2} + \sqrt{(Y_{mn} - a_2)^2 + X_{mn}^2} \right.$$

$$\left. - \sqrt{(Y_{mn} + a_1)^2 + X_{mn}^2} - \sqrt{(Y_{mn} - a_1)^2 + X_{mn}^2} \right]$$

(40)

where $\quad B_m = \dfrac{b_m}{b}, \quad B_n = \dfrac{b_n}{b}, \quad a_1 = \dfrac{\pi}{8}(B_m + B_n) \quad$ and $\quad a_2 = \dfrac{\pi}{8}(B_m - B_n)$

(41)

Note that $\bar{f}_{mn} = 0$ for m = n. We will use Eq. (38) instead of Eq. (39) henceforth.

In the case where there is no vertical and streamwise separation between wings, $X_{mn} = Z_{mn} = 0$, and we get

$$D_{mn} = \frac{\rho \Gamma_m \Gamma_n}{4\pi} \bar{f}_{mn}; \quad \bar{f}_{mn} = \ln \left\{ \frac{[Y_{mn}^2 - (a_n + b_m)^2]}{[Y_{mn}^2 - (a_n - b_m)^2]} \right\}$$

(42)

This expression can be used if one is interested in only the induced drag reduction of the entire flock and not the individuals. The self-induced drag of wing M can be deduced by putting $Y_{mn} = 0$ also:

$$D_{mn} = \frac{\rho \Gamma_m^{max} \Gamma_m^{av}}{4\pi} \bar{f}_{mn}; \quad \bar{f}_{mn} = 2 \ln \left(\frac{b_m + a_m}{b_m - a_m} \right) \tag{43}$$

Note that both average and maximum circulation values of wing M are used. This is simply because the average value obtains if the full span b_m is considered as it is in computing the induced drag due to the horse-shoe vortex of wing M, whose spacing is however $2a_m$ and strength is the maximum circulation. The two are related by $\Gamma_m^{av} = \frac{\pi}{4} \Gamma_m^{max}$ as can be seen from below:

If the lift on a wing is elliptically distributed: $\Gamma(y) = \Gamma^{max} \left(1 - \frac{y^2}{b^2} \right)^{1/2}$, where b is the semi-span, then the average value of the circulation is $\Gamma^{av} = \frac{\pi b \Gamma^{max}}{2} \frac{1}{2b} = \frac{\pi}{4} \Gamma^{max}$. The lift on the wing is $L = \rho V_\infty \int_{-b}^{b} \Gamma(y) dy = \rho V_\infty \left(\frac{\pi b \Gamma^{max}}{2} \right) = \rho V_\infty (2a\Gamma^{max}) = \rho V_\infty (2b\Gamma^{av})$ so that the lifting wing can be represented by a horse-shoe vortex of strength Γ^{max} with spacing $2a$, as is the usual custom with $a = \frac{\pi}{4} b$. However, for calculating D_{mn}, we have a choice of using Γ^{max} and replacing b_m in earlier equations by a_m, in which case D_{mn} in Eq. (42) becomes singular, or using Γ^{av} as we have done. Eq. (43) therefore gives for the self-induced drag of wing M:

$$D_I = \frac{\rho \Gamma_m^{max} \Gamma_m^{av}}{2\pi} \ln \left(\frac{4 + \pi}{4 - \pi} \right) = 0.337 \rho \Gamma_m^{max} \Gamma_m^{av} \tag{44}$$

which is much smaller than that for an elliptically loaded wing, which can be deduced as follows:

$$D_I = \frac{1}{\pi q} \left(\frac{W}{2b} \right)^2 \text{ and } W = \rho V_\infty \Gamma^{max} (2a) \rightarrow \frac{\Gamma^{max}}{V_\infty} = \frac{2}{\pi q} \left(\frac{W}{2b} \right) = \frac{2W}{\pi \rho V_\infty^2 b} = \frac{S C_L}{\pi b}$$

so that $\quad D_I = \frac{\pi}{8} \rho (\Gamma^{max})^2 = \frac{2}{\pi} \rho (\Gamma^{av})^2 = \frac{1}{2} \rho \Gamma^{max} \Gamma^{av} \tag{45}$

The underestimate is due to replacing the bound vortex of varying strength and the vortex sheet behind the elliptically loaded wing by a horse-shoe vortex of strength Γ^{max} and spacing $2a$, a gross approximation at best. So we will use Eq. (45) for the induced drag of an isolated wing so that the fractional reduction in induced drag of wing M due to wing N is:

$$\frac{D_{mn}}{D_{Im}} = \frac{\left(\dfrac{\rho \Gamma_m^{av} \Gamma_n^{max}}{4\pi} \bar{f}_{mn}\right)}{\left(\dfrac{\rho \Gamma_m^{av} \Gamma_n^{max}}{2}\right)} = \frac{1}{2\pi}\left(\frac{\Gamma_n^{max}}{\Gamma_m^{max}}\right)\bar{f}_{mn} = \frac{1}{2\pi}\frac{\left(\frac{W}{2b}\right)_n}{\left(\frac{W}{2b}\right)_m}\bar{f}_{mn} \qquad (46)$$

Thus only the span loading of the two wings and the geometrical factor $\bar{f}_{mn}$ given by Eqs. (37), (39) or (42) need to be considered. For a formation of N_w wings, the total fractional reduction in the induced drag of wing M is

$$\left(\frac{\Delta D_I}{D_I}\right)_m = \frac{1}{D_{Im}}\sum_{n=1}^{N_w} D_{mn} = \frac{1}{2\pi\left(\frac{W}{2b}\right)_m}\sum_{n=1}^{N_w}\left(\frac{W}{2b}\right)_b \bar{f}_{mn} \qquad (47)$$

Note that $\bar{f}_{mn} = 0$. The fractional reduction in total drag can be obtained by multiplying Eq. (47) by the ratio $\left(\frac{C_{D_i}}{C_D}\right)_m$ so that, since power expended is proportional to the drag (velocity being constant), the fractional power reduction of wing M due to other wings in the formation is:

$$\eta_m = \frac{\Delta P_m}{P_m} = \left(\frac{\Delta D}{D}\right)_m = \left(\frac{C_{D_i}}{C_D}\right)_m\left(\frac{\Delta D_I}{D_I}\right)_m = \frac{1}{2\pi\left(\frac{W}{2b}\right)_m}\left(\frac{C_{D_i}}{C_D}\right)_m\sum_{n=1}^{N_w}\left(\frac{W}{2b}\right)_n \bar{f}_{mn} \quad (48)$$

The advantage of writing it this way is that one can substitute for $\left(\frac{C_{D_i}}{C_D}\right)$ a value of $\frac{1}{2}$ for flight at maximum range and $\frac{3}{4}$ for flight at minimum power.

The total reduction in power demand of the entire formation is $\Delta P = \sum_{m=1}^{N_w}\Delta P_m$. The power demand of the wings in the formation if each flew solo is $P = \sum_{m=1}^{N_w} P_m$ so that the fractional power reduction of the whole formation

$$\eta = \frac{\Delta P}{P} = \frac{\sum_{m=1}^{N_w}\Delta P_m}{\sum_{m=1}^{N_w} P_m} = \frac{\sum_{m=1}^{N_w}\Delta D_m}{\sum_{m=1}^{N_w} D_m} = \frac{\sum_{m=1}^{N_w}\eta_m D_m}{\sum_{m=1}^{N_w} D_m}. \qquad (49)$$

It can be shown that this is a consequence of Munk's stagger theorem, which states that "The total induced drag of a collection of lifting surfaces is unaltered by any translation of the wings in the flight direction as long as the circulation of every wing is unchanged." This means that η is independent of X_{mn}, even though η_m is a strong function of X_{mn} of the wings in the formation. Therefore, a flock of birds in line abreast formation will have the same induced drag (and hence the same total power saved) if the formation changed to a V formation or an echelon formation. Thus staggering in the flight direction does not change the total induced drag and hence the total power reduction or power reduction per bird in the flock. However, staggering causes redistribution of the induced drag, and hence the power saving of an individual bird depends on the formation and its location in that formation. In a V-formation, the leading bird and its immediate followers, and the ones toward

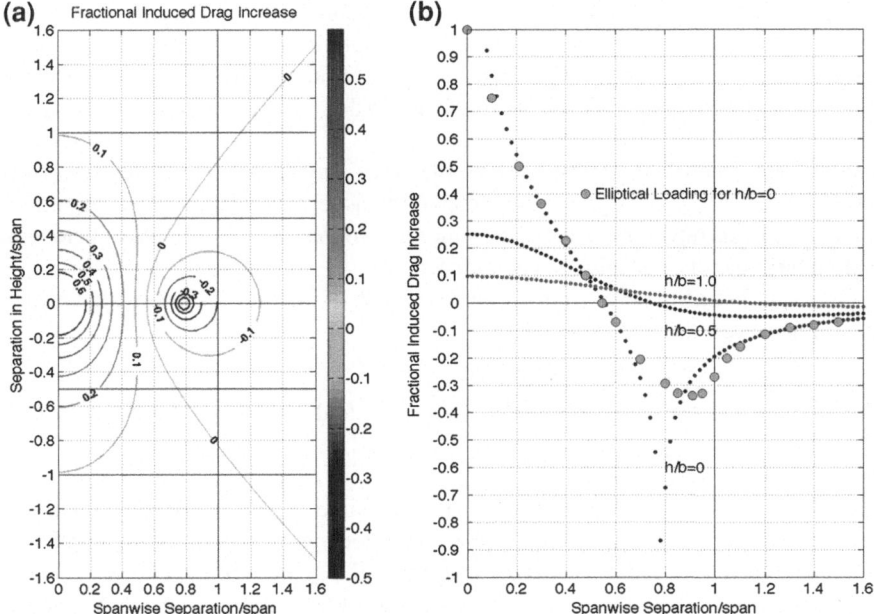

Fig. 15 Fractional induced drag increase during the formation flight of two identical birds

the ends of the V benefit less from formation flight than the ones in the middle of the arms of V. In a line-abreast formation, the birds toward the ends benefit less and the same holds for echelon formation.

Equation (48) can be used to determine the power saving of each bird in formation flight of any shape—V, W, J, echelon or abreast—formation. There is no restriction either on the bird characteristics such as span, weight and wing loading, which can assume arbitrary values.

The special case of just two identical birds in formation flight is instructive. Figure 15a shows the fractional increase in induced drag of each bird for various spanwise and vertical separations. The beneficial location is to the right of the zero contour line. The maximum benefit occurs when the two wings are slightly overlapping with $Y_{mn} \sim 0.8$. To the left of the zero contour line, there is drag penalty since the induced drag increases over the solo flight value. If the two birds flew exactly one behind the other, the induced drag would double if the vertical separation is zero. However, with vertical spacing beyond roughly one span, the induced drag increase becomes less than 10%. Note that if the lift were to reverse on wing B (as in a mirrored wing in ground effect), the areas of beneficial and detrimental induced drag change would switch, with the regions to the right of the zero contour being detrimental instead of beneficial. This fact can be used to compute the induced drag changes, when the formation flight is in ground effect and each wing is mirrored in the ground (lift vector is reversed).

Figure 15b shows the fractional induced drag change as a function of the spanwise separation for various vertical separations. For flight in the same horizontal plane, a slight overlap in the wings gives the maximum reduction. However, flight with one bird immediately behind the other doubles the induced drag. Exact solution from King and Gopalarathnam (2005) for elliptically loaded wings flying at the same horizontal level is shown as green filled circles. Clearly, the horse-shoe vortex model tends to work poorly for $0.7 < Y_{mn} < 1.1$. Another noteworthy aspect is that the vertical separation of $h/b > 0.5$ cuts down the induced drag increase so that by flying above or below the leading bird, the drag penalty can be reduced.

One simple sub case is the formation flight of identical birds with uniform spacing between them. In this case,

$$\eta_m = \left(\frac{\Delta C_D}{C_D}\right)_m = \frac{1}{2\pi}\left(\frac{C_{Di}}{C_D}\right)_m \sum_{n=1}^{N_W} \bar{f}_{mn} \quad \text{and} \quad \eta = \frac{1}{N_W}\sum_{m=1}^{N_W} \eta_m \qquad (50)$$

Consider a flock of identical birds with the same spanwise spacing. Figure 16 shows the fractional power saving of the entire flock as a function of the number of birds for different ratios of spanwise spacing to the span (s/b) for two well known flight conditions: 1. Minimum power expended (blue solid lines) and 2. Maximum range (black dotted lines). Recall that the power saving does not depend on the shape of the formation, although the distribution of fractional power saving among birds depends very much on the shape. Clearly, the power saving decreases rapidly with the number of birds and the benefit of formation flight does not increase much beyond flock strength of around 30 or so. The spanwise spacing is rather critical. A tight formation, no matter what the shape, leads to large power savings. The best benefit is for wing overlap corresponding to $\frac{s}{b} = \frac{\pi}{4} - 1$ and the benefit can be as much as 75% for minimum power flight (and 50% for maximum range). At this point, the induced drag of the flock is equal to that of a single bird! However, this configuration is hard to maintain without birds drifting to the inside of the horseshoe vortex of the adjacent bird and hence coming under the influence of its downwash instead of its upwash. It is probably easier to maintain an overlap corresponding to s/b = −0.1 or 0. Even then, the power savings are significant. Beyond s/b values of around 1 or so, the power saving becomes rather minimal. As a consequence, the flock must maintain a tight formation, irrespective of the shape of the formation, whether abreast, echelon or V. Naturally, from the point of view of visual contact with the birds ahead, either an echelon or V formation is highly desirable. These are indeed the most commonly observed formations. Of course, all the birds in the formation must be at the same level to realize the maximum benefit of formation flight, although again from the point of view of visual contact, it helps to fly slightly above the birds ahead in the formation.

It is also noteworthy that as the flight speed increases, C_{D0}/C_{Di} increases reflecting the fact that the induced drag decreases with speed while the form and skin friction drag increase. Therefore, the power saving comes down as flight

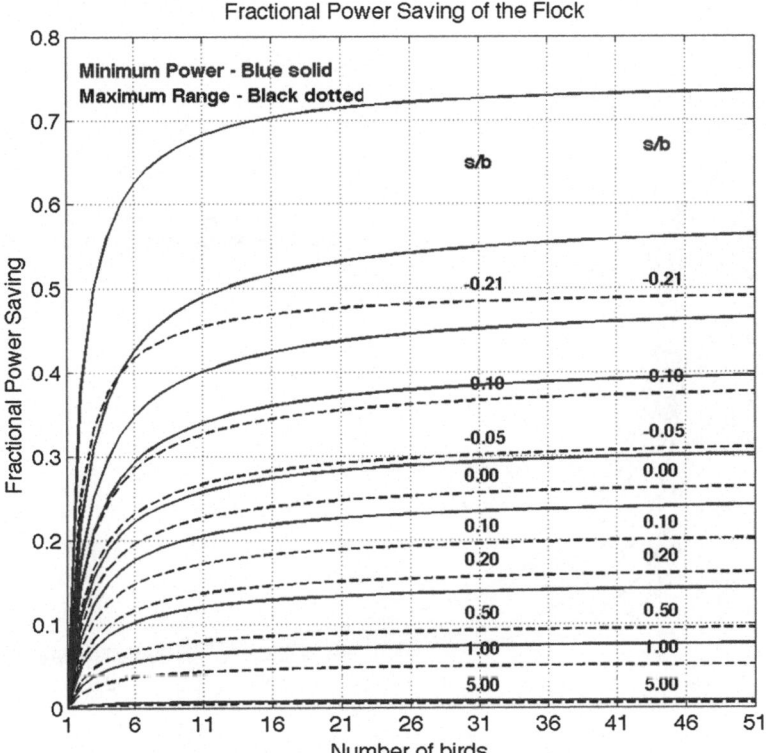

Fig. 16 The fractional power saving of the entire flock as a function of number of birds for different values of s/b. *Blue curves* correspond to minimum power, for which the flight speed is lower than that corresponding to maximum range. *Black curves* correspond to maximum range

speed increases. This can be seen from Fig. 16, which shows that for the same s/b values, the black curves corresponding to maximum range lie below the red lines corresponding to minimum power. The flight speed for maximum range is 31.6% higher than that for minimum power. The benefits of formation flight are best realized if the flight speeds are low. This may encourage the formation to fly at speeds corresponding to minimum power expenditure by the flock as a whole.

An alternative way to present power saving due to formation flight is to present the fractional power saving as a function of wing spacing s/b for different number of birds. Figure 17 does so for flock speeds corresponding to the velocity for maximum range (black dotted) and minimum power (blue solid) of solo flight. Clearly, even formation of 2–3 birds reaps significant power savings. It is not at all unusual to see formations of as few as 2–5. An additional noteworthy fact is that as the number of birds increases, the fractional power saving tends to asymptote beyond a value of around 25 birds in the flock.

However, the fractional power saving of individual birds in the formation depends on its location in the formation and the shape of the formation. This can

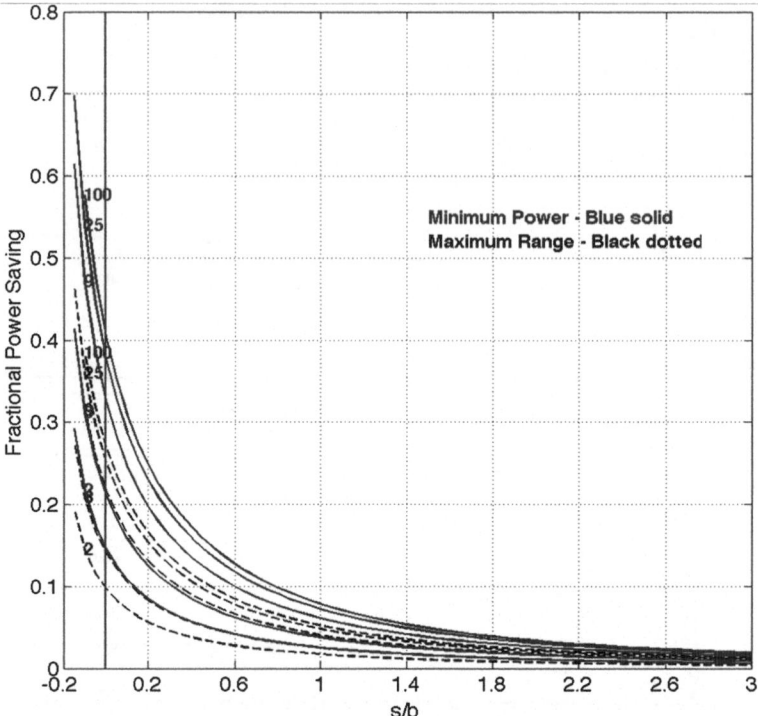

Fig. 17 The fractional power saving of the entire flock as a function of s/b for different values of number of birds. *Blue solid curves* correspond to minimum power, for which the flight speed is lower than that corresponding to maximum range. *Black dotted curves* correspond to maximum range

be seen from Fig. 18, which shows the values for each bird of the formation of 21 birds, for abreast, echelon and V formations and flight speed corresponding to the minimum power requirement of individual birds. The fractional power saving is 0.283 for the entire flock, but the saving for an individual bird depends on its position in the formation. The saving is particularly small for the leading bird in both echelon and V formations. This is simply because the upwash decreases significantly ahead of the horseshoe vortex and the leading bird, because of its location, does not benefit from the trailing vortices of any bird in the formation. In the particular example shown, the fractional power saving of the leading bird is only 0.08 for V formation and 0.04 for echelon formation.

This does not apply to abreast formation since then every bird benefits from the trailing vortices of the rest of the flock. Noticeably the trailing bird in either the V or echelon formation does not suffer significantly (the fractional power savings are 0.270 and 0.271, respectively compared to the overall flock saving of 0.283). The birds in the middle of both formations benefit more. Consequently, it helps to be in the middle of these formations and it is not unusual for birds joining the formation to try to do so in the middle! On the other hand, the birds at each end of the abreast

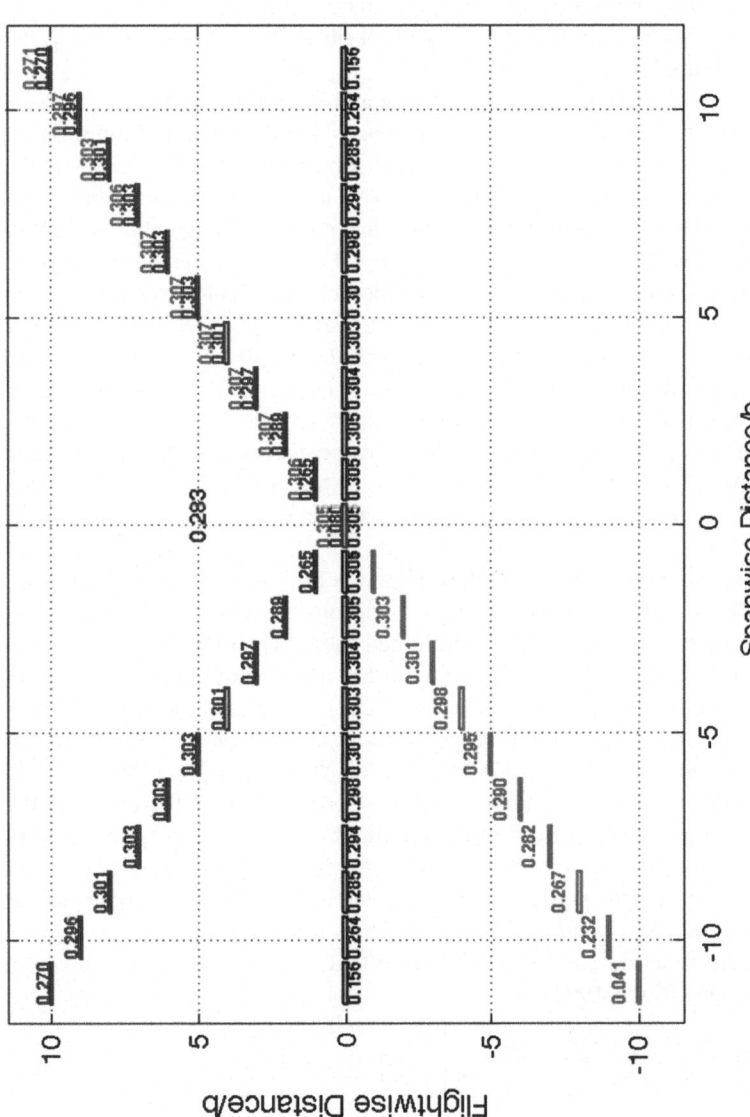

Fig. 18 The fractional power saving of individual birds in the formation for a flock consisting of 21 birds. The fractional power saving is 0.283 for the entire flock. But the saving for an individual bird depends on its position in the formation

formation do not derive as much benefit (only 0.156 compared to the overall flock saving of 0.283) as the rest of the flock. Since the benefit increases toward the center of the flock and becomes maximum there (0.305), during long flights, the formation might tend toward V formation because the birds with lesser benefit might not be able to keep up with the abreast formation. It is in fact interesting to note that a slight deformation of the abreast formation into a U formation leads to roughly equal distribution of power saving to all birds in the flock, including the leading bird (Hummel 1983).

The fact that the leading bird in both echelon and V formations derives the least (but nonzero) benefit from formation flight indicates that during long flights, either the formation must be led by a strong bird or another bird must switch places with the leading bird. In fact, switching has been observed to happen. If the formation is traveling at a speed that is considerably less than that corresponding to the maximum continuous power available from the flight muscles, it is possible that the leading bird can keep its position in the formation for considerable time as long as it is not too weak. There are indications that migrating birds travel at flight speeds considerably higher than those corresponding to the maximum range, simply because the available power output of the flight muscles permits them to. However, the fractional power saving of the flock and hence the individual birds decreases as the flight speed increases. As shown earlier, the migration distance depends only on the aerodynamic efficiency (L/D) of the bird and the fat (and muscle) fraction of the total body mass converted into mechanical energy. Thus, there is an incentive to travel at maximum possible L/D ratio. While the birds are certainly capable of recognizing when they are expending minimum power and when they are at the limit of their muscular power, they might not be able to recognize the flight conditions corresponding to maximum L/D, unless of course evolution has conditioned them. This is unlikely since the attainable range is such a complex and unknowable function of parameters such as the wind speed and direction along the entire migration path and the flight altitude.

The flight speed for maximum range for formation flight is less than that corresponding to an individual bird. However, this flight speed is rather close to the solo flight speed for minimum power. This is fortuitous since the leader of the flock derives minimum benefit from formation flight and is therefore quite likely to fly at a speed corresponding to solo minimum power condition, forcing others in the formation to fly at that speed, which however is close (but not equal) to the speed for optimum range condition for the flock as a whole.

The geometric function can be simplified so that one can write, for the whole flock (Kshatria and Blake 1992),

$$\frac{\Delta C_{Di}}{C_{Di}} = \frac{4}{\pi^2 N_W} \sum_{n=1}^{N_W-1} \sum_{m=n+1}^{N_W} \ln\left[1 - \left(\frac{\frac{\pi}{4}}{|m-n|\left(1+\frac{s}{b}\right)}\right)^2\right]. \tag{51}$$

where s is the spacing between wing tips and b is the wing span. Therefore the total fractional power saving of the flock is

$$\boxed{\frac{\Delta P}{P} = \frac{|\Delta C_D|}{C_D} = \frac{|\Delta C_{Di}|}{C_{D0} + C_{Di}} = \frac{\dfrac{|\Delta C_{Di}|}{C_{Di}}}{1 + \dfrac{C_{D0}}{C_{Di}}}.} \tag{52}$$

Note that $\frac{C_{D0}}{C_{Di}} = 1$ for maximum range and $1/3$ for minimum power. In the limit $N_W \to \infty$,

$$\frac{\Delta C_{Di}}{C_{Di}} = \frac{4}{\pi^2} \ln \left[\sin \left(\frac{\pi^2/4}{1 + \frac{S}{b}} \right) \bigg/ \left(\frac{\pi^2/4}{1 + \frac{S}{b}} \right) \right] \tag{53}$$

For the special case of $S = \frac{\pi}{4} - 1$, the trailing vortices of adjacent wings overlap and cancel so that the induced drag of the formation becomes the induced drag of a wing of span $N_w b$! The induced drag of the flock is equal to the induced drag of a single bird flying solo. Therefore

$$\frac{|\Delta C_{Di}|}{C_{Di}} = \frac{1}{N_w} - 1 \text{ and the fractional power saved is}$$

$$\frac{\Delta P}{P} = \frac{|\Delta C_D|}{C_D} = \frac{\left(1 - \dfrac{1}{N_W}\right)}{\left(1 + \dfrac{C_{D0}}{C_{Di}}\right)} \tag{54}$$

5 Great Flight Diagrams

Finally, heavier-than-air flight, both in nature by birds, bats and insects, and by man-made aircraft, requires wings to produce the lift needed to sustain flight. Consequently, it is not unrealistic to expect certain universal scaling to exist. Tennekes (2009) has popularized the concept that the wing loading, defined as the weight of the flying object W (unit is N) divided by the wing area S (unit is m^2) sustaining the flight, is the most universal attribute of flight, whether it is natural flying objects such as birds and butterflies, or man-made airplanes. When the weight of the flying object W is plotted against the *wing loading WL = W/S* (unit is N/m^2), the data points tend to fall around a universal line for weights ranging from 10^{-5} N corresponding to a fruit fly to 5.6×10^6 N corresponding to the heaviest civilian airliner Airbus A380 (his Fig. 2). This is nearly 12 orders of magnitude change in weight. Such a correlation is quite impressive, especially given the fact that wing loading appears prominently in considerations of flight speed of birds and planes, and landing and takeoff speeds of aircraft. Tennekes (2009) calls the plot the Great Flight Diagram.

Figure 19 shows W plotted against W/S for data tabulated in Appendix A, which includes those from Tennekes (2009). Unlike in Tennekes (2009), the data

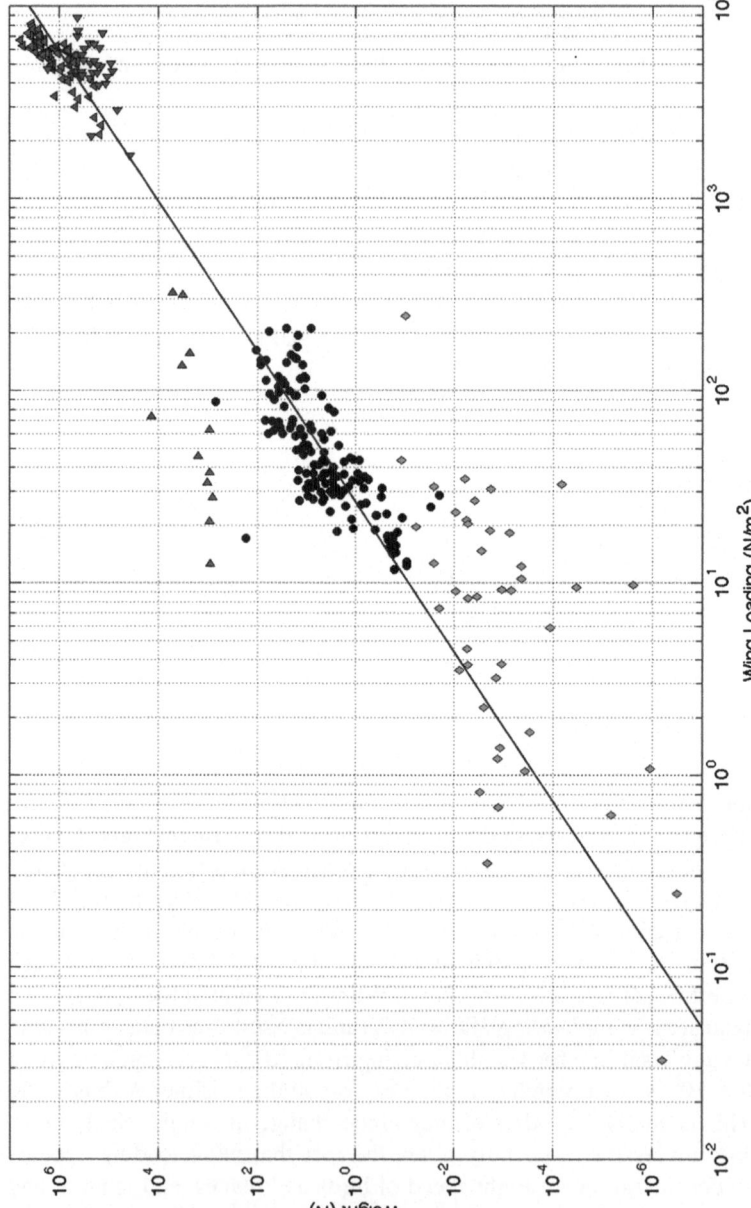

Fig. 19 The Great Flight Diagram with weight W plotted against wing loading WL = W/S, updated from *Tennekes* (2009). *Circles* correspond to birds, with different colors for different sources of data, *squares* to bats, *diamonds* to insects, and *vertical triangles* to aircraft. *Right-pointing triangles* denote human-powered aircraft and *left-pointing triangles* to fighter aircraft. *Blue line* is the regression line in his Fig. 2

includes bats and a large number of insects, and is more comprehensive. While there exists an approximate correlation between W and W/S as in Tennekes (2009), a closer examination shows considerable scatter and significant departures from the regression line, especially for insects. The scatter is likely due to the fact that the data points correspond to flight objects with disparate aspect ratios. Also, wing loading is a parameter that determines the performance of the wing so far as its cruise, landing and takeoff speed characteristics are concerned, and not necessarily the overall wing design. Consequently, we see significant departures of insect wings (green diamonds) as well as gliders and human-powered aircraft (right-pointing red triangles) from the regression line. Given the different structural design and the material of bird and insect wings, this is not surprising. The same applies to the wings of human-powered aircraft, which are low weight, low speed designs by necessity. Military fighter aircraft (leftt-pointing red triangles), whose design emphasis is more on maneuverability and performance in combat, rather than low landing and takeoff speeds, also depart significantly from the regression line, whereas birds (circles) and bats (squares), and transport aircraft (vertical red triangles) pretty much cluster around the regression line.

There are other parameters that influence the design of wings, both natural and man-made. One such parameter is the *span loading SL = W/b* (unit is N/m), which determines the induced drag of the wing. As mentioned earlier, the induced drag can be thought of as the penalty for production of lift by wings of finite span, the induced drag being zero for a wing of infinite span. The induced drag is not a function of the wing loading; the ratio of the induced drag to the parasitic drag is, this ratio also being a function of the aspect ratio. This suggests that for flying objects with widely varying aspect ratios, the parameter that better optimizes the wing design might be the span loading. In fact, Templin (2000) suggests that it is the span loading that is more important to flight and not the wing loading or the aspect ratio, even though wing loading is of importance to takeoff and landing. However, he uses wing span b normalized by the body length l, which he defines as $l = (W/35)^{1/3}$ for his analyses of the flight of birds and bats. It is therefore useful to replot the Great Flight Diagram using the span loading instead of the wing loading as the independent parameter. Figure 20 shows such a plot of W vs SL, from data tabulated in Appendixes A–H, including those from Tennekes (2009). Note that wing span is a quantity more easily determined for nature's flyers than the wing area and consequently, the data on wing span is more abundant and explains the vastly larger number of data points in Fig. 20 compared to Fig. 19.

The correlation between the weight and span loading is tighter *over 8 orders of magnitude change* in SL, than that between the weight and wing loading *over 5 orders of magnitude change* in WL, as can be seen by comparing Figs. 19 and 20. The scatter is significantly diminished. The span loading data for gliders and very light craft follow a similar slope to that of birds, since the emphasis in the wing design is on light weight. Birds of course have been evolutionarily fine-tuned to be very light fliers. The aircraft data also fall roughly on the same line.

Figure 21 shows a plot of the weight W versus span loading SL for just the nature's fliers: birds and bats. Note that the wingspan is readily measured, whereas

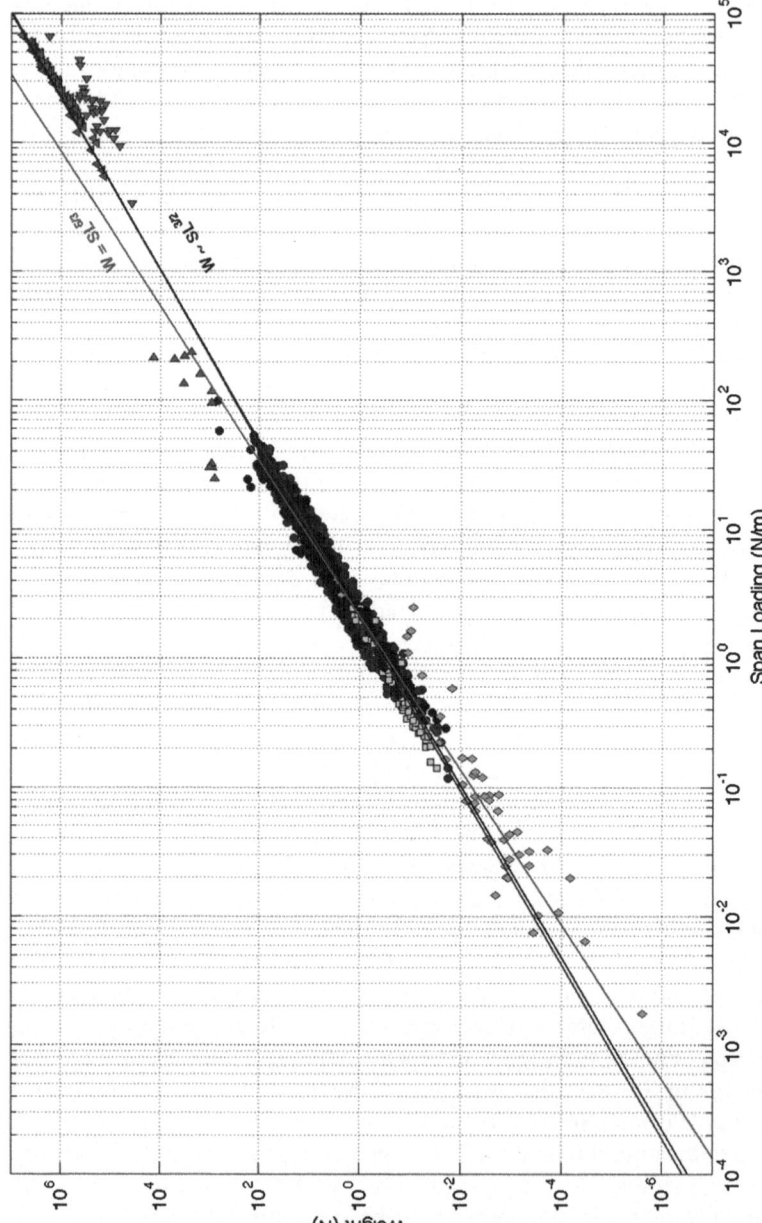

Fig. 20 The second Great Flight Diagram resulting from plotting weight W versus span loading SL = W/b from data in Appendix A–H. *Circles correspond to birds, with different colors for different sources of data, squares to bats, diamonds to insects, and vertical triangles to aircraft. Right-pointing triangles denote human-powered aircraft and left-pointing triangles to fighter aircraft. Blue line is the least square fit line*

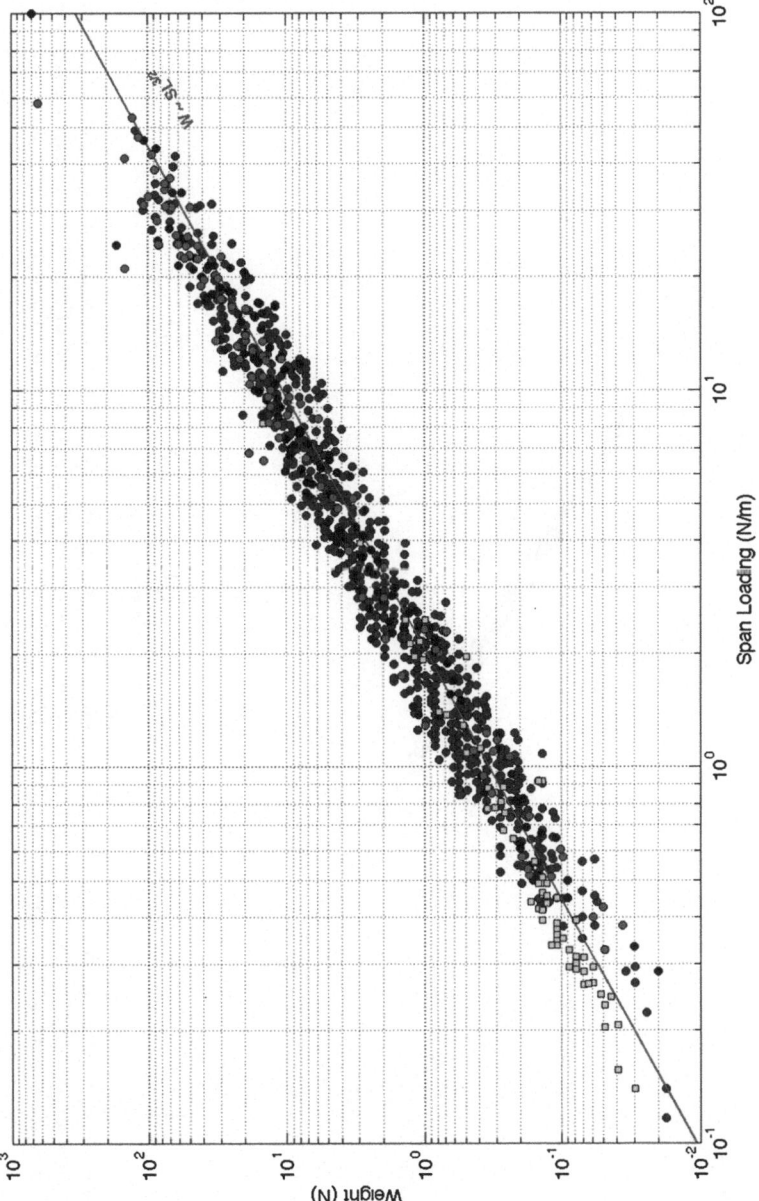

Fig. 21 The second Great Flight Diagram for birds and bats. Weight W is plotted against span loading SL = W/b for all available data, which are tabulated in Appendices A to H. *Circles* correspond to birds, with different colors for different sources of data and *squares* to bats. Note the tight correlation over four orders of magnitude in weight!

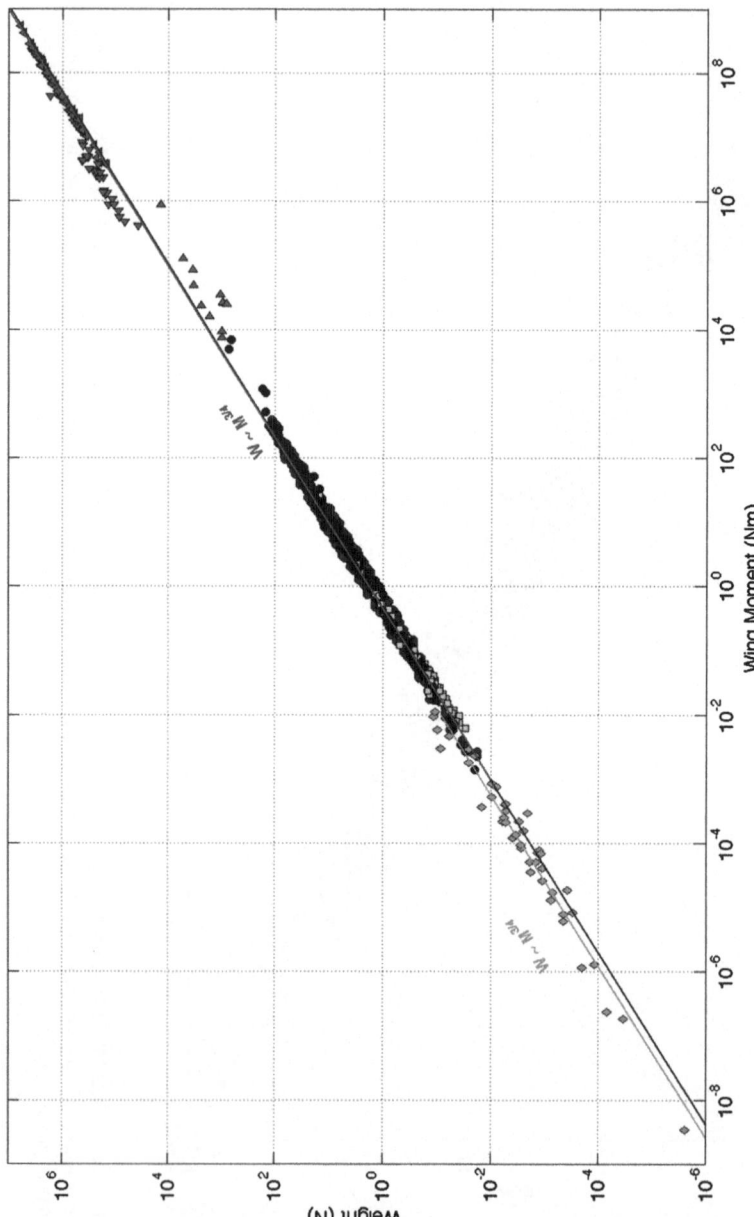

Fig. 22 The third Great Flight Diagram resulting from plotting weight W vs the wing moment WM = Wb from data in Appendix A–H. *Circles correspond to birds, with different colors for different sources of data, squares to bats, diamonds to insects, and vertical triangles to aircraft. Right-pointing triangles denote human-powered aircraft and left-pointing triangles to fighter aircraft. Blue line is the least square fit to all the data*

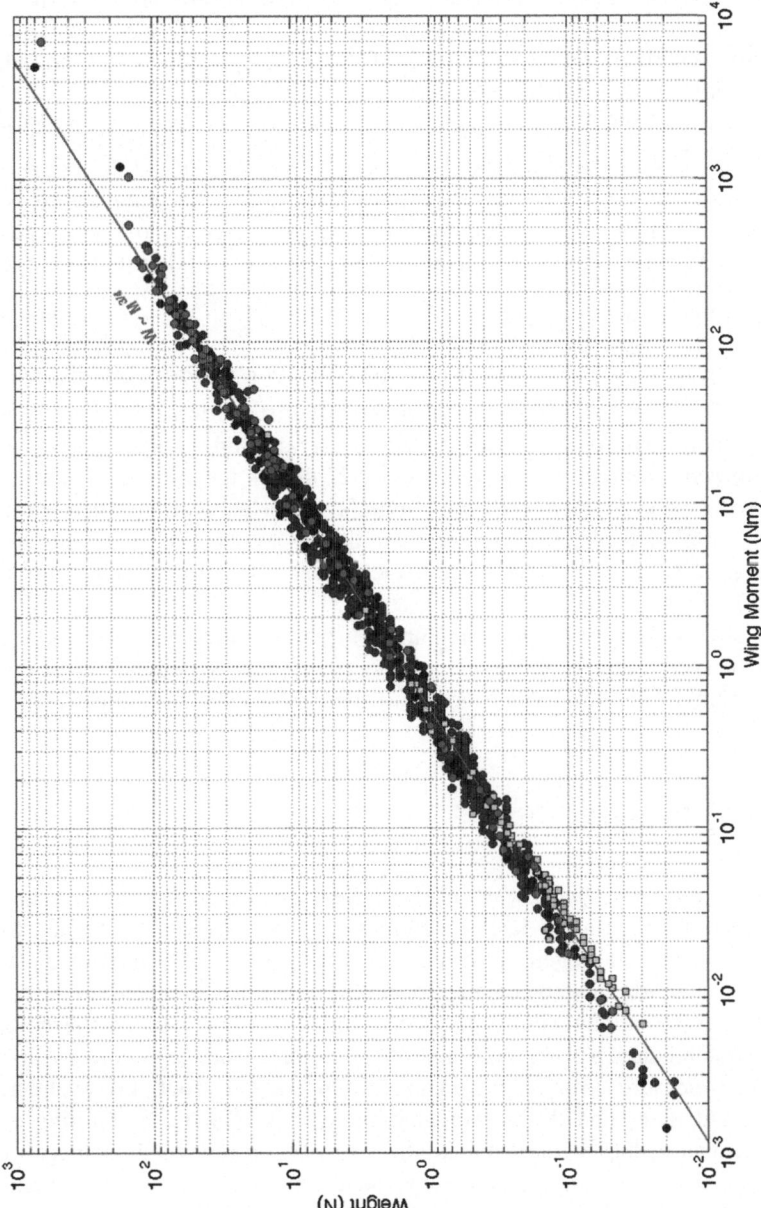

Fig. 23 The third Great Flight Diagram for birds and bats. Weight W is plotted against wing moment $WM = Wb$ for all available data, which are tabulated in Appendices A to H. *Circles* correspond to birds, with different colors for different sources of data and *squares* to bats. Note the tight correlation over six orders of magnitude in wing moment!

the wing area is harder to discern for birds, bats and insects. Consequently, there are far more data on wingspan available for birds and bats than wing area. These data are tabulated in Appendix B, C and D for birds and E for bats, and plotted in Fig. 21. It is noteworthy that all these data cluster beautifully around $W \sim SL^{3/2}$ line and the correlation is tight! The black circles correspond to Tennekes data for birds (Appendix A), blue circles to data for birds listed in Appendices B, C and D, and the red squares to data for bats (Appendix E).

The third Great Flight Diagram parameter is the moment at the wing root equal to $M = \int_0^{b/2} L(y)y\,dy$, where $L(y)$ is the local lift on the wing and y is the spanwise distance. This moment determines how strong the wing has to be and therefore its weight. The higher this moment, the heavier the wing has to be. The exact value of this moment depends on the lift distribution. For the canonical elliptic lift distribution $L(y) = L_c \sqrt{1 - \left(\frac{2y}{b}\right)^2}$, $M = \frac{L_c b^2}{12}$. Since $W = \int_{-b/2}^{b/2} L(y)dy = \frac{\pi b L_c}{4}$, $M = Wb/(3\pi) \sim 0.106\,Wb$. For uniform lift distribution, $M = 0.125\,Wb$. Thus no matter what the lift distribution, the wing root moment is proportional to Wb. Thus the third Great Flight Diagram can be based on *wing root moment WM = Wb*. Figure 22 shows the weight W plotted against the wing root moment Wb for nature's flyers as well as man-made aircraft. The correlation is even tighter than the W versus SL plot *over 16 orders of magnitude change* in WM from the tiniest insect to the superjumbo A380 (compare Figs. 21 and 22)!

Figure 23 shows the third flight diagram for just the nature's flyers. Once again, the tight correlation is self-evident. The correlation is better than in the second Great Flight Diagram, which is in turn better than the first Great Flight Diagram.

6 Concluding Remarks

While migration on wings has always been of great interest to humanity, it has assumed increased importance lately. Migrating birds have the potential to quickly spread harmful pathogens worldwide. The possibility of H5N1 avian flu virus becoming a worldwide pandemic (once the virus has mutated to infect humans) with the potential to inflict huge causalities has underscored the need to better understand and predict bird migration patterns and behavior, so that effective countermeasures can be put in place. On another front, the eventual depletion of fossil fuels, and the phenomenal changes in climate the burning of fossil fuels can cause, has focused mankind's attention on renewable sources of energy such as winds. However, care must be taken to make sure that wind turbine farms are located so as not to interfere with, hurt or kill migrating birds.

Fortunately, modern technology gives us the capability to study migration much better than has been possible in the past, as the example of the bar-tailed godwit cited above shows. New technologies are coming on stream for monitoring bird migration (Fiedler 2009) and proven techniques such as radar tracking (Nebuloni et al. 2008) and satellite tracking are being increasingly useful. Hopefully, in the coming decades, we will be able to completely decipher this fascinating phenomenon of migration on wings.

Finally, it is important for us to realize that migration is in itself an arduous task and not all the birds, which embark on seasonal migrations complete it (Obviously, enough do to preserve and propagate the species). Migrating birds encounter numerous perils along the way: adverse weather, hunters and predators, to name a few. At the end of their journey, the birds are not only tired, but also very famished, especially if they have encountered adverse weather conditions along the way. This highlights the importance of establishing and preserving bird sanctuaries along migration paths, where the birds can rest and rebuild their fat reserves undisturbed by hunters before taking off on their journey again. This is the least we can do to preserve this epic phenomenon of migration on wings for future generations to observe and enjoy!

References

T. Alerstam, *Bird Migration* (Cambridge University Press, Cambridge, 1990)

T. Alerstam, A. Hedenstrom, The development of bird migration theory. J. Avian Biol. **29**, 343–369 (1998)

J. Elphick (ed.), *Atlas of Bird Migration*. (Firefly Books, Buffalo, 2009), p. 176

W. Fiedler, New technologies for monitoring bird migration and behaviour. Ringing Migr. **24**, 175–179 (2009)

C.H. Greenewalt, The flight of birds. Trans Am Philos Soc New Ser. **65**, 1–67 (1975)

J.J.L. Higdon, S. Corrsin, Induced drag of a bird flock. Am. Nat. **112**, 727–744 (1978)

D. Hummel, Aerodynamic aspects of formation flight in birds. J. Theor. Biol. **104**, 321–347 (1983)

R.M. King, A. Gopalarathnam, Ideal aerodynamics of ground effect and formation flight. J. Aircr. **42**, 1188–1199 (2005)

M. Kshatriya, R.W. Blake, Theoretical model of the optimum flock size of birds flying in formation. J. Theor. Biol. **157**, 135–174 (1992)

F. Liechti, Birds: blowin'by the wind? J. Ornithol. **147**, 202–211 (2006)

R. Nebuloni, C. Capsoni, V. Vigorita, Quantifying bird migration by a high-resolution weather radar. IEEE J. Geosci. Remote Sens. **46**, 1867–1875 (2008)

C.J. Pennycuick, The mechanics of bird migration. Ibis **111**, 525–556 (1969)

C.J. Pennycuick, *Bird Flight Performance: A Practical Calculation Manual* (Oxford University Press, New York, 1989)

C.J. Pennycuick, *Modelling the Flying Bird* (Academic Press, Amsterdam, 2008), p. 480

C.J. Pennycuick, P.F. Battley, Burning the engine: a time-marching computation of fat and protein consumption in a 5420 km nonstop flight by great knots, Calidris tenuirostris. OIKOS **103**, 323–332 (2003)

J. Perrin, *Winged Migration* (Chronicle Books, California, 2003), p. 272

D.P. Raymer, *Aircraft Design: A Conceptual Approach*, 4th edn. (AIAA Press, Washington, 2006), p. 838

Rozell N. Bar-tailed godwit goes the distance. Article #1876, Alaska Science Forum, 10 Oct 2007

W. Shyy, M. Berg, D. Ljungqvist, Glapping flight and flexible wings for biological and micro air vehicles. Prog. Aerosp. Sci. **35**, 455–505 (1999)

R.J. Templin, The spectrum of animal flight: insects to pterosaurs. Prog. Aerosp. Sci. **36**, 393–436 (2000)

H. Tennekes, *The Simple Science of Flight* (MIT Press, London, 2009), p. 201

Appendix A
Data on Birds from "The Simple Science of Flight by H. Tennekes, 2009, MIT Press" (Reproduced with Permission)

Number	Weight (N)	Wing area (m^2)	Span (m)	L/D	Speed (m/s)	Name
1	0.020	0.0007	0.07	0.0	0.0	Bee-hummingbird
2	0.030	0.0012	0.09	0.0	0.0	Rubythroat-hummingbird
3	0.090	0.0073	0.18	0.0	10.6	Coal-Tit
4	0.090	0.0070	0.20	4.0	0.0	Magnolia-Warbler
5	0.110	0.0050	0.17	4.0	0.0	House-Wren
6	0.140	0.0076	0.21	0.0	10.6	Siskin
7	0.150	0.0096	0.27	0.0	14.3	Sand-martin
8	0.150	0.0104	0.29	0.0	9.7	House-martin
9	0.160	0.0136	0.32	0.0	10.0	Barn-swallow
10	0.170	0.0120	0.31	8.0	0.0	Barn-swallow
11	0.170	0.0100	0.33	0.0	0.0	Storm-petrel
12	0.170	0.0100	0.32	10.0	0.0	Chimney-swift-16
13	0.180	0.0103	0.26	0.0	12.7	Yellow-wagtail
14	0.290	0.0103	0.26	4.0	12.7	Yellow-wagtail
15	0.190	0.0109	0.23	0.0	13.6	Great-tit
16	0.200	0.0130	0.33	0.0	0.0	Barn-swallow
17	0.200	0.0130	0.32	8.0	0.0	Tree-swallow
18	0.210	0.0126	0.26	0.0	14.1	White-wagtail
19	0.220	0.0126	0.27	0.0	12.7	Tree-pipit
20	0.230	0.0100	0.24	4.0	0.0	Orchard-oriole
21	0.280	0.0090	0.23	4.0	0.0	Couse-sparrow
22	0.380	0.0168	0.40	0.0	9.7	Swift
23	0.360	0.0160	0.42	13.0	0.0	Swift
24	0.390	0.0207	0.35	0.0	15.1	Skylark
25	0.540	0.0156	0.36	0.0	15.3	Dunlin
26	0.610	0.0233	0.36	0.0	13.8	Redwing
27	0.640	0.0179	0.41	0.0	19.5	Ringed-plover

(continued)

L. Kantha, *Migration on Wings*, SpringerBriefs in Applied Sciences and Technology, DOI: 10.1007/978-3-642-27925-6, © The Author(s) 2012

(continued)

Number	Weight (N)	Wing area (m²)	Span (m)	L/D	Speed (m/s)	Name
28	0.680	0.0218	0.36	0.0	11.0	Song-thrush
29	0.780	0.0304	0.57	0.0	12.6	Alpine-swift
30	0.820	0.0240	0.38	7.0	0.0	American-robin
31	0.830	0.0244	0.38	7.0	16.2	Starling
32	0.830	0.0190	0.41	9.0	0.0	Purple-martin-36
33	0.890	0.0240	0.38	7.0	0.0	Blue-Jay
34	1.100	0.0571	0.80	0.0	10.9	Arctic-tern
35	1.110	0.0252	0.47	0.0	14.9	Turnstone
36	1.140	0.0333	0.44	0.0	11.9	Mistle-thrush
37	1.200	0.0560	0.83	14.0	0.0	Common-tern
38	1.280	0.0286	0.50	14.0	20.1	Red-knot
39	1.400	0.0440	0.60	9.0	0.0	Merlin
40	1.620	0.0644	0.54	0.0	12.9	Eurasian-jay
41	1.700	0.0560	0.75	10.0	0.0	Hobby
42	1.700	0.0460	0.00	0.0	0.0	Dove-prion
43	1.740	0.0406	0.61	0.0	12.3	Greenshank
44	1.800	0.0600	0.74	10.0	0.0	Kestrel
45	2.030	0.0708	0.73	0.0	10.1	Kestrel
46	2.190	0.0420	0.62	0.0	17.9	Grey-plover
47	2.190	0.0744	0.75	0.0	17.9	Northern-lapwing
48	2.300	0.0750	0.00	0.0	0.0	Blackheaded-gull
49	2.380	0.0667	0.73	0.0	11.3	Hobby
50	2.400	0.1300	1.10	11.0	0.0	Montagus-harrier
51	2.450	0.0684	0.65	12.0	12.5	Jackdaw
52	2.500	0.0800	0.75	0.0	11.3	Sparrow-hawk
53	2.500	0.0800	0.95	0.0	0.0	Franklins-gull
54	2.700	0.0350	0.56	0.0	0.0	Puffin
55	2.770	0.0768	0.67	0.0	0.0	Sparrow-hawk
56	2.800	0.0700	0.78	0.0	0.0	Rock-dove
57	2.900	0.0750	0.80	8.0	0.0	Rock-dove
58	2.830	0.0976	0.97	0.0	11.9	Blackheaded-gull
59	2.970	0.0891	1.01	0.0	13.6	Longheaded-skua
60	3.000	0.0890	0.00	0.0	0.0	Black-skimmer
61	3.180	0.0520	0.73	14.0	18.3	Bartailed-godwit
62	4.900	0.0520	0.73	12.0	18.3	Bartailed-godwit
63	3.300	0.1100	1.03	0.0	0.0	Laughing-gull
64	3.300	0.1400	1.00	10.0	0.0	Hen-harrier-male
65	3.480	0.0428	0.59	0.0	19.7	Common-teal
66	3.870	0.1040	0.95	0.0	12.8	Elenoras-falcon
67	4.080	0.0953	0.96	0.0	13.1	Kittiwake
68	4.110	0.1246	1.11	0.0	13.4	Common-gull
69	4.300	0.0770	0.88	0.0	0.0	Cape-pigeon
70	4.300	0.0900	0.71	9.0	0.0	Coopers-hawk

(continued)

(continued)

Number	Weight (N)	Wing area (m²)	Span (m)	L/D	Speed (m/s)	Name
71	4.330	0.1570	1.10	0.0	9.1	Hen-harrier
72	4.380	0.1180	1.06	0.0	13.8	Arctic-skua
73	4.700	0.1080	1.15	14.0	0.0	Royal-tern
74	4.700	0.1760	1.15	11.0	0.0	Hen-harrier-female
75	4.880	0.1380	0.91	0.0	13.5	Rook
76	4.910	0.0824	0.75	8.0	16.3	Wood-pigeon
77	5.000	0.1680	1.12	8.0	0.0	Barn-Owl
78	5.660	0.1380	0.91	10.0	13.5	Carrion-crow
79	6.530	0.2040	1.16	10.0	11.2	Marsh-harrier
80	6.800	0.2000	1.30	0.0	0.0	Marsh-harrier
81	7.000	0.1700	0.97	9.0	0.0	Goshawk-male
82	7.100	0.2600	1.25	0.0	0.0	Goshawk
83	7.190	0.1930	1.34	14.0	13.1	Blackbacked-gull
84	7.780	0.2470	1.26	0.0	12.5	Honey-buzzard
85	7.890	0.1257	1.02	10.0	12.1	Peregrine-falcon
86	8.000	0.1660	1.02	11.0	0.0	Redshouldered-hawk
87	8.000	0.0380	0.68	0.0	0.0	Razorbill
88	8.100	0.1300	1.06	0.0	0.0	Peregrine-falcon
89	8.200	0.1240	0.00	0.0	0.0	Fulmar
90	8.850	0.2690	1.24	11.0	11.6	Buzzard
91	9.400	0.1810	1.40	0.0	0.0	Herring-gull
92	9.430	0.3320	1.35	0.0	10.5	Roghlegged-buzzard
93	10.000	0.2700	1.35	0.0	0.0	Common-buzzard
94	10.120	0.3250	1.66	0.0	12.0	Red-kite
95	10.240	0.0879	0.90	0.0	20.6	Pintail
96	10.820	0.1060	0.88	0.0	18.5	Mallard
97	11.000	0.0930	0.90	0.0	0.0	Mallard
98	11.000	0.2090	1.22	11.0	0.0	Redtailed-hawk
99	11.420	0.1970	1.34	14.0	12.6	Herring-gull
100	11.500	0.2470	1.21	0.0	14.3	Raven
101	12.000	0.2400	1.15	10.0	0.0	Goshawk-female
102	12.000	0.0880	0.72	4.0	0.0	Pheasant
103	13.000	0.2600	1.45	12.0	0.0	Osprey-male
104	13.100	0.1130	1.01	12.0	17.7	Brent-goose
105	13.500	0.2140	0.00	0.0	0.0	Great-skua
106	13.900	0.5150	1.47	0.0	11.7	Spotted-eagle
107	14.400	0.3720	1.73	0.0	12.5	Grey-heron
108	14.900	0.0770	0.93	0.0	19.7	Goosander
109	15.000	0.4400	1.75	11.0	0.0	Turkey-Vulture
110	15.100	0.0890	1.04	0.0	18.6	Redthroated-diver
111	15.800	0.3200	1.60	0.0	13.3	Osprey
112	15.900	0.1080	0.93	0.0	16.0	King-eider
113	16.000	0.1700	1.35	12.0	19.0	Barnacle-goose

(continued)

(continued)

Number	Weight (N)	Wing area (m^2)	Span (m)	L/D	Speed (m/s)	Name
114	16.700	0.2880	1.67	0.0	0.0	Blackbeaked-gull
115	17.000	0.1800	1.40	0.0	0.0	Whitefronted-goose
116	17.100	0.1150	1.08	12.0	17.0	Barnacle-goose
117	19.200	0.2720	0.00	0.0	13.7	Blackbilled-gull
118	20.000	0.3000	1.60	13.0	0.0	Osprey-female
119	20.200	0.1310	0.98	0.0	17.9	Common-eider
120	21.000	0.3300	1.38	10.0	0.0	Black-vulture
121	22.300	0.2240	1.40	10.0	15.2	Cormorant
122	25.400	0.1200	1.20	0.0	19.3	Blackthroated-diver
123	25.800	0.1840	1.41	0.0	16.1	Whitefronted-goose
124	27.000	0.2500	1.85	0.0	0.0	Gannett
125	28.000	0.3400	2.18	20.0	0.0	Sooty-albatross
126	30.000	0.5000	2.20	0.0	0.0	Brown-pelican
127	30.400	0.2680	1.62	0.0	17.3	Bean-goose
128	33.300	0.3080	1.55	0.0	17.1	Greylag-goose
129	34.300	0.5330	1.91	10.0	16.0	White-stork
130	36.300	0.3720	1.69	0.0	16.7	Canada-goose
131	37.000	0.5400	2.12	0.0	11.9	Golden-eagle
132	38.000	0.3200	2.20	0.0	0.0	Blackbrowed-albatross
133	38.000	0.3600	2.16	20.0	0.0	Blackbrowed-albatross
134	40.700	0.5970	2.03	14.0	0.0	Golden-eagle
135	45.000	0.5000	2.00	0.0	0.0	Sandhill-crane
136	47.000	0.7600	2.24	15.0	0.0	Bald-eagle
137	50.000	0.7200	2.20	14.0	0.0	Whitetailed-eagle
138	56.100	0.5860	2.22	0.0	15.0	Common-crane
139	57.000	0.2800	1.70	14.0	0.0	Canada-goose
140	60.000	1.0000	2.80	0.0	0.0	White-pelican
141	66.400	0.4610	1.98	0.0	18.5	Tundra-swan
142	68.000	0.6000	2.20	0.0	0.0	Whooping-crane
143	70.000	1.0000	2.60	15.0	0.0	Griffon-Vulture
144	85.000	0.6200	3.40	25.0	0.0	Wandering-albatross
145	86.900	0.6050	1.98	0.0	18.5	Whooper-swan
146	106.000	0.6500	2.30	10.0	16.2	Mute-swan
147	170.000	10.0000	7.00	0.0	0.0	Pteranodon
148	700.000	8.0000	7.00	0.0	0.0	Argentavis-magnificens

Appendix B
Data on Aircraft Including Data from "The Simple Science of Flight by H. Tennekes, 2009" (Reproduced with Permission)

Number	Weight (N)	Wing area (m²)	Span (m)	L/D	Speed (m/s)	Name
1	939.000	70.6000	29.30	0.0	0.0	Gossamer-Condor-Human-Plane
2	975.000	44.0000	28.60	0.0	0.0	Gossamer-Albatross-Human-Plane
3	840.000	28.2000	32.00	0.0	0.0	NihonMowe20-Human-Plane
4	1090.000	30.8000	34.10	38.0	0.0	Daedalus-Human-Plane
5	14500.000	186.2000	64.01	22.0	45.0	Solar-Impulse
6	1000.000	15.0000	9.93	8.0	0.0	Hang-glider
7	1000.000	25.0000	8.00	4.0	0.0	Parawing
8	1700.000	35.0000	10.00	4.0	0.0	Powered-parawing
9	2500.000	15.0000	10.00	8.0	0.0	Ultralight
10	3500.000	10.5000	15.00	40.0	0.0	Sailplane
11	3600.000	25.0000	25.00	35.0	0.0	Icare-Solarplane
12	5500.000	16.0000	25.00	60.0	0.0	Open-class-Sailplane
13	1659000.000	260.0000	44.84	0.0	0.0	Airbus-A300-600
14	1419740.000	219.0000	43.90	0.0	0.0	Airbus-A310
15	590000.000	122.6000	34.09	0.0	231.0	Airbus-A318
16	758870.000	122.6000	34.09	0.0	237.0	Airbus-A319
17	770200.000	122.6000	34.09	0.0	231.0	Airbus-A320
18	834610.000	122.6000	34.09	0.0	231.0	Airbus-A321
19	2399500.000	361.6000	60.30	0.0	239.0	Airbus-A330-200
20	2750000.000	361.6000	60.30	0.0	239.0	Airbus-A340-200-300
21	3800000.000	439.4000	63.45	0.0	239.0	Airbus-A340-500-600
22	3680000.000	439.0000	63.00	0.0	0.0	Airbus-340-600
23	2650000.000	443.0000	64.80	0.0	250.0	Airbus-A350-900
24	5600000.000	845.0000	80.00	19.0	0.0	Airbus-A380-800
25	2500000.000	345.0000	64.49	0.0	206.0	Antonov-AN-22

(continued)

L. Kantha, *Migration on Wings*, SpringerBriefs in Applied Sciences and Technology, 51
DOI: 10.1007/978-3-642-27925-6, © The Author(s) 2012

(continued)

Number	Weight (N)	Wing area (m²)	Span (m)	L/D	Speed (m/s)	Name
26	4050000.000	628.0000	73.30	0.0	236.0	Antonov-AN-124
27	6400000.000	905.0000	88.40	0.0	222.0	Antonov-AN-225
28	470000.000	95.8000	28.50	0.0	242.0	BAC-111-500
29	491900.000	91.0500	28.35	0.0	252.0	Boeing-737-100
30	523900.000	91.0500	28.35	0.0	251.0	Boeing-737-200
31	612500.000	105.4000	28.88	0.0	252.0	Boeing-737-300
32	680500.000	105.4000	28.89	0.0	253.0	Boeing-737-400
33	605500.000	105.4000	28.89	0.0	257.0	Boeing-737-500
34	655440.000	125.0000	34.31	0.0	257.0	Boeing-737-600
35	700800.000	125.0000	34.31	0.0	257.0	Boeing-737-700
36	790160.000	125.0000	34.31	15.0	257.0	Boeing-737-800-900
37	1510000.000	283.0000	44.00	0.0	0.0	Boeing-707-320B
38	1720000.000	283.0000	48.00	0.0	0.0	Boeing-767-300
39	3401000.000	511.0000	59.64	0.0	266.0	Boeing-747-100
40	3778000.000	511.0000	59.64	0.0	266.0	Boeing-747-200
41	3156000.000	511.0000	59.64	0.0	261.0	Boeing-747-SP
42	3401000.000	511.0000	59.64	0.0	252.0	Boeing-747-300
43	3968900.000	524.9000	64.44	16.0	252.0	Boeing-747-400
44	4400000.000	570.0000	69.00	0.0	0.0	Boeing-747-8
45	1156800.000	185.2000	38.05	0.0	237.0	Boeing-757-200
46	1224700.000	185.2000	38.05	0.0	237.0	Boeing-757-300
47	1428810.000	283.4000	47.57	0.0	237.0	Boeing-767-200
48	1597550.000	283.4000	47.57	0.0	237.0	Boeing-767-300
49	2041700.000	290.7000	51.99	0.0	237.0	Boeing-767-400
50	2472100.000	427.8000	60.93	0.0	237.0	Boeing-777-200
51	2975600.000	427.8000	60.93	0.0	237.0	Boeing-777-300
52	2970000.000	427.8000	60.93	20.0	0.0	Boeing-777-200ER
53	3510000.000	435.0000	65.00	20.0	0.0	Boeing-777-300ER
54	2450000.000	370.0000	60.00	0.0	0.0	Boeing-787-9
55	240000.000	55.0000	21.00	0.0	0.0	Bombardier-CL600
56	740000.000	0.0000	35.00	0.0	0.0	Embraer-KC390
57	560000.000	146.7000	34.30	0.0	0.0	Caravelle-10B
58	1474150.000	266.5000	43.41	0.0	259.0	Douglas-DC-8
59	210000.000	51.0000	20.00	0.0	0.0	Embraer-ERJ145
60	430000.000	94.0000	28.00	0.0	0.0	Fokker-F100
61	640000.000	140.0000	37.90	0.0	180.0	Illyushin-18D
62	1650000.000	279.5000	43.20	0.0	255.0	Illyushin-62M
63	2150000.000	300.0000	48.06	0.0	255.0	Illyushin-86
64	2600000.000	350.0000	60.11	0.0	255.0	Illyushin-96
65	1950450.000	320.0000	47.34	0.0	247.0	Lockheed-L-1011
66	2630850.000	367.7000	50.40	0.0	252.0	MD-DC10
67	411400.000	86.8000	27.25	0.0	245.0	MD-DC9-10
68	548850.000	93.0000	28.47	0.0	245.0	MD-DC9-30-40-50

(continued)

(continued)

Number	Weight (N)	Wing area (m²)	Span (m)	L/D	Speed (m/s)	Name
69	2733140.000	338.9000	51.66	0.0	245.0	MD-MD11
70	707600.000	112.3000	32.87	0.0	230.0	MD-MD90
71	388200.000	84.0000	27.80	0.0	230.0	Sukhoi-SSJ-100
72	494500.000	84.0000	27.80	0.0	230.0	Sukhoi-SSJ-100
73	445000.000	127.3000	29.00	0.0	230.0	Tupolev-TU-134
74	900000.000	201.5000	37.55	0.0	250.0	Tupolev-TU-154
75	1000000.000	201.5000	37.55	0.0	250.0	Tupolev-TU-154M
76	946000.000	182.4000	41.80	0.0	250.0	Tupolev-TU-204
77	1107500.000	182.4000	41.80	0.0	250.0	Tupolev-TU-204-220
78	461000.000	83.2000	29.77	0.0	230.0	Tupolev-TU-334-100
79	544200.000	100.0000	32.61	0.0	230.0	Tupolev-TU-334-100D
80	1880000.000	310.0000	50.10	0.0	220.0	Tupolev-TU95
81	1415230.000	264.8000	44.55	0.0	230.0	Vickers-VC10
82	160000.000	70.0000	25.00	0.0	130.0	Yakolev-YAK40
83	570000.000	150.0000	34.88	0.0	230.0	Yakolev-YAK42
84	1300000.000	356.0000	42.40	0.0	220.0	Airbus-A400M
85	210000.000	74.9800	29.20	0.0	139.0	Antonov-AN-24
86	270000.000	75.0000	29.20	0.0	131.0	Antonov-AN-24
87	2500000.000	345.0000	64.49	0.0	206.0	Antonov-AN-22
88	2213500.000	371.6000	56.39	20.0	272.0	Boeing-B52G
89	2567300.000	371.6000	56.39	20.0	280.0	Boeing-B52G
90	2653500.000	353.0000	51.74	0.0	222.0	Boeing-C17A
91	151000.000	59.1000	25.81	0.0	128.0	CASA-CN235
92	1570000.000	300.0000	50.50	0.0	0.0	Illyushin-76D
93	2100000.000	300.0000	50.50	0.0	0.0	Illyushin-78M
94	703000.000	162.1000	40.40	0.0	164.0	Lockheed-C130H
95	3810000.000	576.0000	67.89	0.0	259.0	Lockheed-C5B
96	510000.000	160.0000	40.00	0.0	139.0	Transall-C160
97	254000.000	56.5000	13.05	0.0	0.0	Boeing-F15A
98	254000.000	37.2000	11.43	0.0	0.0	Boeing-FA18C
99	120000.000	28.4000	9.60	0.0	0.0	Dassault-SuperEtender
100	162000.000	25.0000	8.40	0.0	0.0	Dassault-Mirage-F1C
101	170000.000	41.0000	9.13	0.0	0.0	Dassault-Mirage-2000
102	245000.000	45.7000	10.80	0.0	0.0	Dassault-Rafale
103	226800.000	47.0000	17.53	0.0	0.0	Fairchild-A10A
104	453000.000	61.1000	19.20	15.8	0.0	GeneralDynamics-F111-Spread
105	453000.000	48.8000	9.75	0.0	0.0	GeneralDynamics-F111-Swept
106	150000.000	28.9000	9.45	0.0	0.0	GeneralDynamics-F16
107	337000.000	52.5000	19.54	0.0	0.0	Grumman-F14-spread
108	337000.000	52.5000	10.15	0.0	0.0	Grumman-F14-swept
109	89000.000	18.2000	6.68	0.0	0.0	Lockheed-F104A

(continued)

(continued)

Number	Weight (N)	Wing area (m^2)	Span (m)	L/D	Speed (m/s)	Name
110	141000.000	18.2000	6.68	0.0	0.0	Lockheed-F104S
111	150000.000	28.9000	9.45	0.0	0.0	Lockheed-F16
112	238000.000	105.9000	13.20	0.0	0.0	Lockheed-F117A
113	380000.000	78.0000	13.56	0.0	0.0	Lockheed-F22
114	111000.000	24.2000	8.38	0.0	0.0	McDonnel-Douglas-A4
115	71000.000	23.0000	7.15	0.0	0.0	Mikoyan-Gurevich-MIG-21F
116	180000.000	37.4000	13.97	0.0	0.0	Mikoyan-Gurevich-MIG-23-Spread
117	180000.000	34.2000	8.10	0.0	0.0	Mikoyan-Gurevich-MIG-23-Swept
118	367000.000	61.4000	14.01	0.0	0.0	Mikoyan-Gurevich-MIG-25P
119	210000.000	38.0000	11.40	0.0	0.0	Mikoyan-Gurevich-MIG-29
120	93000.000	17.3000	8.13	0.0	0.0	Northrop-F5E
121	437500.000	55.2000	17.64	0.0	0.0	Sukhoi-24-spread
122	437500.000	55.2000	10.37	0.0	0.0	Sukhoi-24-swept
123	205000.000	30.1000	14.36	0.0	0.0	Sukhoi-25TM
124	345000.000	62.0000	14.70	0.0	0.0	Sukhoi-27
125	370000.000	78.8000	14.20	0.0	0.0	Sukhoi-T50
126	40000.000	22.5000	11.20	0.0	0.0	Spitfire
127	1800000.000	358.0000	25.60	6.0	0.0	Concorde

Appendix C
Data on Birds from "Atlas of Bird Migration by J. Elphick, 2007" (Reproduced with Permission) (Metric Units if Ind = 1, British Units if Ind=0)

Number	Mass range (kg/lb)		Span range (m/inch)		Migration range (km/mile)		Ind	Name
1	0.3700	0.4250	0.610	0.800	0.0	11200.0	1	Bluewinged-Teal
2	0.8150	0.9300	0.660	0.800	500.0	5000.0	1	Oldsquaw
3	0.1100	0.1200	0.530	0.580	0.0	6000.0	1	American-Kestrel
4	0.9100	1.0700	1.280	1.280	6000.0	12000.0	1	Swaison-Hawk
5	1.5000	1.5000	1.700	1.700	0.0	6000.0	1	Turkey-Vulture
6	6.4000	7.3000	2.000	2.300	4000.0	4000.0	1	Whooping-Crane
7	2.9000	5.8000	1.800	2.100	0.0	4000.0	1	sandhill-Crane
8	0.0900	0.1000	0.460	0.560	0.0	10000.0	1	Killdeer
9	0.2300	0.3400	0.890	0.940	3000.0	8000.0	1	Franklin-Gull
10	0.4500	0.4500	1.080	1.140	0.0	8000.0	1	Royal-tern
11	0.0620	0.0620	0.540	0.610	4000.0	11000.0	1	Common-Nighthawk
12	0.0510	0.0550	0.460	0.500	500.0	6000.0	1	Whippoor-will
13	0.0520	0.0520	0.360	0.400	0.0	4000.0	1	Common-Poorwill
14	0.0240	0.0240	0.310	0.320	3000.0	10000.0	1	Chimney-Swift
15	0.0200	0.0200	0.310	0.340	1000.0	5000.0	1	Tree-Swallow
16	0.0190	0.0190	0.320	0.340	2500.0	11000.0	1	Barn-Swallow
17	0.0030	0.0035	0.100	0.120	0.0	6000.0	1	Rubythrated-Hummingbird
18	0.0025	0.0030	0.110	0.110	2000.0	5000.0	1	Calliope-Hummingbird
19	0.0500	0.0500	0.350	0.400	0.0	3500.0	1	Yellowbilled-Sapsucker
20	0.0140	0.0140	0.250	0.270	0.0	4000.0	1	Vermilion-Flycatcher
21	0.0430	0.0430	0.360	0.390	2000.0	4000.0	1	Scissortailed-Flycatcher
22	0.0770	0.0770	0.370	0.420	0.0	6400.0	1	American-Robin
23	0.0140	0.0240	0.240	0.270	4000.0	10000.0	1	Redeyed-Vireo

(continued)

L. Kantha, *Migration on Wings*, SpringerBriefs in Applied Sciences and Technology, DOI: 10.1007/978-3-642-27925-6, © The Author(s) 2012

(continued)

Number	Mass range (kg/lb)		Span range (m/inch)		Migration range (km/mile)		Ind	Name
24	0.0290	0.0290	0.280	0.300	1000.0	7000.0	1	Scarlet-Tanager
25	0.0110	0.0110	0.200	0.240	4000.0	8000.0	1	Blackpoll-Warbler
26	0.0140	0.0140	0.210	0.210	1900.0	1900.0	1	Kirtland-Warbler
27	0.0180	0.0260	0.240	0.250	0.0	4000.0	1	Darkeyed-Junco
28	0.0260	0.0260	0.230	0.250	0.0	4500.0	1	Whitethroated-Sparrow
29	0.0800	0.0850	0.390	0.390	0.0	1000.0	1	European-Starling
30	0.0390	0.0490	0.290	0.350	0.0	2000.0	1	Brownheaded-Cowbird
31	0.0410	0.0640	0.300	0.370	0.0	2500.0	1	Redwinged-Blackbird
32	0.0370	0.0470	0.250	0.320	8000.0	11000.0	1	Bobolink
33	2.1900	3.7500	38.000	42.000	1100.0	2500.0	0	Western-Greebe
34	0.5600	0.8700	22.000	23.000	300.0	3000.0	0	Eared-Greebe
35	3.5000	4.6200	49.000	54.000	300.0	3400.0	0	Double-Crested-Cormorant
36	0.8100	1.2500	41.000	49.000	500.0	4200.0	0	American-Bittern
37	0.2500	0.3750	15.000	23.000	600.0	2500.0	0	Least-Bittern
38	1.1250	1.9400	26.000	29.000	125.0	1250.0	0	Wood-Duck
39	2.5600	3.0000	33.000	37.000	185.0	2175.0	0	AmericanBlack-Duck
40	1.9000	3.5000	32.000	36.000	300.0	3400.0	0	Canvasback
41	1.7500	1.9000	27.000	31.000	600.0	5600.0	0	Lesser-Scaup
42	0.7500	1.0000	21.000	24.000	300.0	3400.0	0	Bufflehead
43	0.9000	1.6200	20.000	24.000	600.0	2500.0	0	Ruddy-Duck
44	3.3100	5.5000	55.000	59.000	60.0	600.0	0	Black-Vulture
45	6.6200	15.4400	70.000	90.000	60.0	2800.0	0	Bald-Eagle
46	0.7000	1.5600	39.000	47.000	300.0	3000.0	0	Northern-Harrier
47	0.2500	0.8100	20.000	28.000	60.0	3000.0	0	SharpShinned-Hawk
48	0.4400	1.1250	29.000	37.000	0.0	1250.0	0	Coopers-Hawk
49	0.4400	1.2000	31.000	36.000	1000.0	8700.0	0	BroadWinged-Hawk
50	1.1250	2.6250	47.000	53.000	0.0	1850.0	0	RedTailed-Hawk
51	2.0000	3.9400	51.000	55.000	125.0	2500.0	0	Ferruginous-Hawk
52	1.4400	2.9000	35.000	43.000	0.0	1250.0	0	Prairie-Falcon
53	1.4400	2.9000	35.000	43.000	1250.0	3750.0	0	Peregrine-Falcon
54	0.1250	0.2500	13.000	15.000	1850.0	4350.0	0	Sora
55	0.6250	1.0620	30.000	33.000	300.0	2800.0	0	American-Vocet
56	0.0630	0.1250	16.000	20.000	4000.0	8000.0	0	Semipalmated-Plover
57	0.1250	0.1880	19.000	23.000	370.0	1500.0	0	Mountain-Plover
58	0.2500	0.4370	23.000	28.000	5900.0	8200.0	0	AmericanGolden-Plover
59	0.5620	1.3100	29.000	33.000	900.0	5600.0	0	Marbled-Godwit
60	1.5600	3.0600	31.000	39.000	300.0	3400.0	0	LongBilled-Curlew
61	0.2500	0.3750	25.000	26.000	5900.0	7500.0	0	Upland-Sandpiper

(continued)

(continued)

Number	Mass range (kg/lb)		Span range (m/inch)		Migration range (km/mile)		Ind	Name
62	0.2500	0.5000	27.000	29.000	1500.0	8000.0	0	Greater-Yellowleg
63	0.1250	0.2500	23.000	25.000	1500.0	9300.0	0	Lesser-Yellowleg
64	0.0630	0.1250	21.000	23.000	2175.0	6800.0	0	Solitary-Sandpiper
65	0.0630	0.1250	14.000	15.000	600.0	6800.0	0	Spotted-Sandpiper
66	0.0630	0.1880	15.000	16.000	4475.0	8000.0	0	Wilsons+Phalarope
67	0.1880	0.3130	17.000	20.000	1500.0	5300.0	0	ShortBilled-Dowitcher
68	0.1880	0.3130	19.000	21.000	300.0	8700.0	0	Surfbird
69	0.3750	0.4380	35.000	39.000	300.0	3400.0	0	Bonapartes-Gull
70	1.1250	1.6250	47.000	61.000	60.0	2500.0	0	RingBilled-Gull
71	0.6880	1.1880	43.000	51.000	300.0	4350.0	0	New-Gull
72	1.5630	3.0630	54.000	61.000	60.0	2500.0	0	Herring-Gull
73	0.1880	0.3130	28.000	32.000	125.0	2800.0	0	Forsters-Tern
74	0.0630	0.1250	18.000	21.000	60.0	1850.0	0	Least-Tern
75	0.0630	0.1250	14.000	16.000	2175.0	4700.0	0	BlackBilled-Cuckoo
76	0.0630	0.1880	15.000	18.000	1250.0	5000.0	0	YellowBilled-Cuckoo
77	0.4380	0.9370	35.000	39.000	0.0	3000.0	0	LongEared-Owl
78	0.1250	0.1880	13.000	15.000	300.0	3000.0	0	Flammulated-Owl
79	0.3130	0.5000	20.000	24.000	0.0	1250.0	0	Burrowing-Owl
80	0.1250	0.1410	26.000	27.000	300.0	2500.0	0	ChuckWills-Widow
81	0.0320	0.0630	13.000	15.000	600.0	4350.0	0	Vauxs-Swift
82	0.0040	0.0040	5.000	6.000	125.0	2800.0	0	BroadTailed-Hummingbird
83	0.0040	0.0040	5.000	6.000	300.0	3750.0	0	Rufous-Hummingbird
84	0.2500	0.4370	18.000	20.000	0.0	4700.0	0	Belted-Kingfisher
85	0.0940	0.0940	11.000	12.000	1850.0	6800.0	0	Eastern-Kingbird
86	0.1040	0.1090	11.000	12.000	125.0	1500.0	0	Gray-Kingbird
87	0.1040	0.1090	11.000	12.000	1100.0	3400.0	0	Western-Kingbird
88	0.0940	0.0940	10.000	11.000	155.0	3400.0	0	GreatCrested-Flycatcher
89	0.0630	0.0940	9.000	11.000	1850.0	5600.0	0	OliveSided-Flycatcher
90	0.0940	0.0940	9.000	10.000	1250.0	4475.0	0	EasternWood-Pewee
91	0.0630	0.0940	9.000	10.000	1100.0	4700.0	0	WesternWood-Pewee
92	0.0630	0.0630	9.000	11.000	0.0	3400.0	0	Eastern-Phoebe
93	0.0310	0.0310	9.000	10.000	185.0	1850.0	0	Gray-Flycatcher
94	0.0310	0.0310	8.000	9.000	185.0	2175.0	0	Dusky-Flycatcher
95	0.0310	0.0310	8.000	9.000	300.0	4000.0	0	Hammonds-Flycatcher
96	0.0310	0.0310	7.000	8.000	1250.0	4475.0	0	Least-Flycatcher
97	0.0310	0.0310	8.000	9.000	1100.0	3400.0	0	Acadian-Flycatcher
98	0.0310	0.0310	8.000	9.000	2175.0	6000.0	0	Willow-Flycatcher

(continued)

(continued)

Number	Mass range (kg/lb)		Span range (m/inch)		Migration range (km/mile)		Ind	Name
99	0.0310	0.0310	8.000	9.000	2800.0	4700.0	0	Alder-Flycatcher
100	0.0310	0.0310	7.000	9.000	2800.0	4700.0	0	YellowBellied-Flycatcher
101	0.0310	0.0310	7.000	9.000	300.0	4350.0	0	Western-Flycatcher
102	0.0310	0.0310	10.000	11.000	600.0	5600.0	0	VioletGreen-Swallow
103	0.0470	0.0630	13.000	15.000	600.0	5900.0	0	Purple-Martin
104	0.0470	0.0630	12.000	13.000	1250.0	6800.0	0	Cliff-Swallow
105	0.0160	0.0310	5.000	7.000	60.0	1500.0	0	House-Wren
106	0.0125	0.0250	5.000	6.000	125.0	1850.0	0	Winter-Wren
107	0.0310	0.0310	5.000	8.000	185.0	2500.0	0	Marsh-Wren
108	0.0160	0.0310	5.000	7.000	600.0	2500.0	0	Sedge-Wren
109	0.0160	0.0160	5.000	6.000	185.0	5900.0	0	RubyCrowned-Kinglet
110	0.0160	0.0160	5.000	7.000	60.0	1250.0	0	BlueGray-Gnatcatcher
111	0.0630	0.0790	9.000	12.000	0.0	2500.0	0	Eastern-Bluebird
112	0.0630	0.0790	9.000	12.000	0.0	2800.0	0	Western-Bluebird
113	0.0630	0.0790	10.000	12.000	0.0	3000.0	0	Mountain-Bluebird
114	0.0940	0.1560	11.000	13.000	600.0	3750.0	0	Wood-Thrush
115	0.0470	0.0940	11.000	12.000	2500.0	5000.0	0	Veery
116	0.0470	0.0940	10.000	11.000	1500.0	6800.0	0	Swainsons-Thrush
117	0.0630	0.0630	9.000	11.000	125.0	5000.0	0	Hermit-Thrush
118	0.1090	0.1880	11.000	13.000	300.0	3400.0	0	Northern-Shrike
119	0.0630	0.1250	9.000	10.000	185.0	4000.0	0	Grey-Catbird
120	0.0940	0.1560	9.000	10.000	85.0	2500.0	0	Sage-Thrasher
121	0.1090	0.1720	11.000	12.000	0.0	1500.0	0	Brown-Thrasher
122	0.0470	0.0470	11.000	12.000	1250.0	3000.0	0	Spragues-Pipit
123	0.0780	0.1250	10.000	12.000	0.0	3400.0	0	Cedar-Waxwing
124	0.0160	0.0310	6.000	9.000	435.0	1250.0	0	BlackCapped-Vireo
125	0.0310	0.0470	7.000	9.000	300.0	4000.0	0	YellowThroated-Vireo
126	0.0310	0.0310	7.000	9.000	185.0	2500.0	0	Bells-Vireo
127	0.0310	0.0470	7.000	9.000	155.0	1400.0	0	Gray-Vireo
128	0.0310	0.0470	7.000	9.000	600.0	4700.0	0	BlueHeaded-Vireo
129	0.0470	0.0630	8.000	10.000	1250.0	2800.0	0	BlackWhiskered-Vireo
130	0.0310	0.0470	7.000	9.000	300.0	5600.0	0	Warbling-Vireo
131	0.0310	0.0470	7.000	9.000	2175.0	4700.0	0	Philadelphia-Vireo
132	0.0310	0.0470	7.000	9.000	900.0	3000.0	0	Prothonotary-Vireo
133	0.0160	0.0310	7.000	9.000	1100.0	2500.0	0	BlueWinged-Warbler
134	0.0160	0.0310	7.000	9.000	1500.0	3400.0	0	GoldenWinged-Warbler
135	0.0160	0.0310	7.000	9.000	1850.0	3000.0	0	Tennessee-Warbler

(continued)

(continued)

Number	Mass range (kg/lb)		Span range (m/inch)		Migration range (km/mile)		Ind	Name
136	0.0310	0.0470	7.000	9.000	600.0	5300.0	0	OrangeCrowned-Warbler
137	0.0160	0.0310	7.000	9.000	300.0	3000.0	0	Nashville-Warbler
138	0.0160	0.0310	6.000	8.000	1250.0	2175.0	0	Virginia-Warbler
139	0.0160	0.0310	6.000	8.000	300.0	3000.0	0	Northern-Parula
140	0.0310	0.0310	6.000	8.000	300.0	4700.0	0	BlackandWhite-Warbler
141	0.0310	0.0310	6.000	8.000	1500.0	4000.0	0	BlackThroatedBlue-Warbler
142	0.0160	0.0310	6.000	7.000	2175.0	4350.0	0	Cerulean-Warbler
143	0.0310	0.0310	6.000	7.000	1500.0	4700.0	0	Blackburnian-Warbler
144	0.0160	0.0310	6.000	8.000	1850.0	4000.0	0	ChestnutSided-Warbler
145	0.0160	0.0310	6.000	8.000	1500.0	2500.0	0	CapeMay-Warbler
146	0.0160	0.0310	6.000	8.000	1500.0	4000.0	0	Magnolia-Warbler
147	0.0310	0.0310	7.000	9.000	300.0	6000.0	0	YellowRumped-Warbler
148	0.0310	0.0310	6.000	8.000	155.0	3750.0	0	BlackThroatedGray-Warbler
149	0.0160	0.0310	6.000	8.000	125.0	4000.0	0	Townsends-Warbler
150	0.0310	0.0310	7.000	9.000	185.0	3400.0	0	Hermit-Warbler
151	0.0160	0.0310	6.000	8.000	1100.0	2800.0	0	BlackThroated-Green-Warbler
152	0.0310	0.0310	7.000	9.000	900.0	1500.0	0	GoldenCheeked-Warbler
153	0.0160	0.0310	6.000	8.000	155.0	1250.0	0	Graces-Warbler
154	0.0160	0.0310	6.000	7.000	470.0	2500.0	0	Prairie-Warbler
155	0.0310	0.0310	7.000	9.000	2175.0	3400.0	0	BayBreasted-Warbler
156	0.0310	0.0310	7.000	9.000	0.0	2500.0	0	Pine-Warbler
157	0.0310	0.0310	7.000	9.000	600.0	3000.0	0	Palm-Warbler
158	0.0310	0.0310	6.000	8.000	1500.0	4350.0	0	Mourning-Warbler
159	0.0160	0.0310	6.000	8.000	1250.0	5000.0	0	Macgillivrays-Warbler
160	0.0310	0.0310	6.000	9.000	2500.0	5600.0	0	Connecticut-Warbler
161	0.0160	0.0310	6.000	8.000	1250.0	5000.0	0	Kentucky-Warbler
162	0.0160	0.0310	5.000	7.000	2500.0	5600.0	0	Wilsons-Warbler
163	0.0310	0.0310	6.000	8.000	2500.0	4350.0	0	Hooded-Warbler
164	0.0310	0.0470	9.000	10.000	600.0	4350.0	0	Ovenbird
165	0.0310	0.0470	9.000	10.000	1850.0	4000.0	0	Northern-Waterthrush
166	0.0310	0.0310	6.000	8.000	1500.0	4700.0	0	American-Redstart

(continued)

(continued)

Number	Mass range (kg/lb)		Span range (m/inch)		Migration range (km/mile)		Ind	Name
167	0.0790	0.1250	9.000	12.000	2500.0	5000.0	0	RoseBreasted-Grosbeak
168	0.0310	0.0470	6.000	9.000	470.0	3750.0	0	Indigo-Bunting
169	0.0310	0.0470	6.000	9.000	600.0	3400.0	0	Lazuli-Bunting
170	0.0310	0.0470	6.000	9.000	300.0	3000.0	0	Painted-Bunting
171	0.0630	0.0790	10.000	12.000	300.0	3000.0	0	GreenTailed-Towhee
172	0.0310	0.0470	6.000	9.000	1500.0	2500.0	0	Bairds-Sparrow
173	0.0470	0.0630	8.000	9.000	1500.0	3750.0	0	AmericanTree-Sparrow
174	0.0310	0.0470	6.000	9.000	600.0	2800.0	0	Mccowns-Longspur
175	0.0310	0.0470	8.000	9.000	900.0	4700.0	0	Dickcissel
176	0.1250	0.1880	12.000	14.000	300.0	4000.0	0	Rusty-Blackbird
177	0.1250	0.1560	11.000	13.000	600.0	4350.0	0	Summer-Tanager
178	0.1090	0.1880	11.000	14.000	0.0	1250.0	0	Evening-Grosbeak
179	0.2500	0.4500	0.580	0.640	500.0	5000.0	1	Eurasian-Teal
180	0.7000	1.1000	0.720	0.820	300.0	7500.0	1	Common-Pochard
181	0.5500	0.8000	0.550	0.690	1000.0	4500.0	1	Smew
182	0.0900	0.1500	0.590	0.660	500.0	6500.0	1	Common-Redshank
183	0.1850	0.3000	0.800	0.920	4000.0	6000.0	1	Slenderbilled-Curlew
184	2.5000	3.5000	1.450	1.550	2000.0	6500.0	1	Black-Stork
185	2.3000	4.4000	1.550	1.650	2000.0	10500.0	1	White-Stork
186	0.4500	1.0000	1.350	1.500	4000.0	10000.0	1	European-HoneyBuzzard
187	1.2000	2.0000	1.450	1.700	4000.0	10000.0	1	Osprey
188	0.3400	0.5000	1.100	1.300	6000.0	9500.0	1	Elenora-Falcon
189	0.1300	0.2000	0.660	0.780	7200.0	12000.0	1	Redfooted-Falcon
190	0.0900	0.2000	0.580	0.720	3000.0	9500.0	1	Lesser-Kestrel
191	0.0750	0.1400	0.320	0.350	1000.0	5500.0	1	Common-Quail
192	0.1250	0.2100	0.460	0.530	4300.0	10000.0	1	Corn-Crake
193	5.0000	8.5000	2.300	2.600	4000.0	5000.0	1	Siberianwhite-Crane
194	4.0000	7.0000	2.200	2.450	2000.0	6000.0	1	Common-Crane
195	2.2000	3.0000	1.650	1.850	1500.0	4500.0	1	Demoiselle-Crane
196	0.0900	0.1300	0.720	0.800	5000.0	6000.0	1	Roseate-Tern
197	0.6000	1.0000	1.250	1.550	1000.0	6500.0	1	Blackbacked-Gull
198	0.0550	0.0850	0.640	0.680	3000.0	10500.0	1	Black-Tern
199	0.1000	0.2000	0.470	0.530	1000.0	6000.0	1	European-TurtleDove
200	0.0900	0.1350	0.550	0.600	4500.0	12000.0	1	Common-Cuckoo
201	0.0600	0.0800	0.420	0.460	500.0	5000.0	1	Hoopoe
202	0.0400	0.0650	0.440	0.490	2500.0	10500.0	1	European-Beeeater
203	0.1200	0.1760	0.660	0.730	2500.0	10000.0	1	European-Roller
204	0.0400	0.0450	0.460	0.460	3000.0	12000.0	1	Common-Swift

(continued)

(continued)

Number	Mass range (kg/lb)		Span range (m/inch)		Migration range (km/mile)		Ind	Name
205	0.0125	0.0125	0.280	0.280	1500.0	10000.0	1	sand-Martin
206	0.0190	0.0190	0.320	0.320	0.0	12000.0	1	Barn-Swallow
207	0.0170	0.0280	0.260	0.260	1000.0	8000.0	1	Yellow-Wagtail
208	0.0240	0.0240	0.260	0.260	0.0	1500.0	1	Water-pipit
209	0.0150	0.0240	0.230	0.230	2000.0	6000.0	1	Common-Redstart
210	0.0190	0.0360	0.250	0.250	2500.0	5500.0	1	Common-Nightingale
211	0.0240	0.0410	0.290	0.320	1500.0	16000.0	1	Northern-Wheatear
212	0.0140	0.0210	0.210	0.210	2000.0	9000.0	1	Common-Whitethroat
213	0.0110	0.0110	0.200	0.200	1500.0	6000.0	1	Reed-Warbler
214	0.0080	0.0120	0.200	0.200	4000.0	14000.0	1	Willow-Warbler
215	0.0100	0.0100	0.260	0.260	4000.0	11000.0	1	Redbacked-Shrike
216	0.0125	0.0190	0.230	0.230	2000.0	7000.0	1	Pied-Flycatcher
217	0.0150	0.0280	0.245	0.245	3500.0	13000.0	1	Spotted-Flycatcher
218	0.8750	2.1250	30.000	33.000	155.0	2500.0	0	RedNecked-Greebe
219	0.6890	1.0630	23.000	25.000	155.0	3000.0	0	Horned-Greebe
220	20.0000	24.2500	106.000	141.000	900.0	4000.0	0	White-Pelican
221	0.2500	0.3750	20.000	22.000	1500.0	6800.0	0	Little-Bittern
222	0.5820	0.8120	31.000	36.000	2500.0	5300.0	0	Squacco-Heron
223	0.8750	1.3750	34.000	37.000	300.0	4000.0	0	Little-Egret
224	1.1880	2.6880	47.000	59.000	1850.0	4350.0	0	Purple-Heron
225	1.1880	1.6890	31.000	37.000	2175.0	2800.0	0	Glossy-Ibis
226	2.5630	3.1880	49.000	53.000	300.0	1850.0	0	Hermit-Ibis
227	2.8750	3.7500	45.000	51.000	155.0	2175.0	0	Eurasian-Spoonbill
228	4.4380	8.9370	55.000	68.000	600.0	2800.0	0	Bean-Goose
229	0.8750	2.3750	29.000	33.000	300.0	4700.0	0	Eurasian-Wigeon
230	1.1880	2.8750	31.000	37.000	300.0	5600.0	0	Northern-Pintail
231	0.5630	1.3130	23.000	24.000	2175.0	4350.0	0	Garganey
232	0.7500	2.1880	27.000	33.000	60.0	4700.0	0	Northern-Shoveler
233	1.8130	3.1250	33.000	34.000	60.0	2175.0	0	RedCrested-Pochard
234	1.0000	1.6250	24.000	26.000	60.0	2800.0	0	Ferruginous-Duck
235	0.8750	2.2500	26.000	28.000	60.0	4000.0	0	Tufted-Duck
236	1.1250	1.6250	24.000	27.000	30.0	1250.0	0	Harlequin-Duck
237	1.1250	2.7500	25.000	31.000	300.0	1850.0	0	Common-Goldeneye
238	2.0000	4.7500	32.000	38.000	60.0	1850.0	0	Common-Merganser
239	1.3750	2.0630	62.000	70.000	600.0	2800.0	0	Black-Kite
240	3.5000	4.8130	61.000	70.000	600.0	1850.0	0	Egyptian-Vulture
241	2.9370	5.1250	72.000	76.000	600.0	3750.0	0	ShortToed-Eagle
242	0.5000	1.3130	37.000	47.000	2175.0	7000.0	0	Pallid-Harrier
243	0.5000	1.0000	41.000	47.000	2175.0	7150.0	0	Montagus-Harrier
244	0.3130	0.6250	25.000	29.000	900.0	3000.0	0	Levant-Sparrowhawk
245	2.3750	4.7500	52.000	62.000	1850.0	6400.0	0	LesserSpotted-Eagle

(continued)

(continued)

Number	Mass range (kg/lb)		Span range (m/inch)		Migration range (km/mile)		Ind	Name
246	3.0000	7.0630	61.000	71.000	600.0	4350.0	0	Spotted-Eagle
247	1.1250	2.7500	39.000	47.000	2175.0	6800.0	0	Booted-Eagle
248	0.3130	0.7500	32.000	36.000	600.0	7150.0	0	Eurasian-Hobby
249	0.6880	1.0000	33.000	43.000	1500.0	4000.0	0	Sooty-Falcon
250	0.1250	0.3130	14.000	16.000	1850.0	7000.0	0	Spotted-Crake
251	0.0630	0.1880	13.000	15.000	1250.0	4475.0	0	Little-Crake
252	0.0630	0.1250	12.000	14.000	600.0	4700.0	0	Braillions-Crake
253	13.2500	19.8130	90.000	98.000	600.0	1250.0	0	Manchurian-Crane
254	0.1250	0.1880	23.000	25.000	900.0	3750.0	0	Collared-Pratincole
255	0.1880	0.2500	23.000	26.000	1750.0	6000.0	0	BlackWinged-Pratincole
256	0.0630	0.1040	16.000	18.000	600.0	5300.0	0	Little-Plover
257	0.0940	0.1560	17.000	22.000	1250.0	8000.0	0	Mongolian-Plover
258	0.1880	0.2500	20.000	23.000	900.0	4000.0	0	GreaterSand-Plover
259	0.1250	0.1880	21.000	24.000	1500.0	6700.0	0	Caspian-Plover
260	0.3750	0.5630	27.000	29.000	1250.0	3400.0	0	Sociable-Plover
261	0.0780	0.2040	14.000	16.000	300.0	5000.0	0	Jack-Snipe
262	0.3130	0.5630	18.000	19.000	2175.0	8200.0	0	Great-Snipe
263	0.1880	0.3130	17.000	18.000	1500.0	6000.0	0	Pintail-Snipe
264	0.3750	1.1250	27.000	32.000	900.0	7800.0	0	BlackTailed-Godwit
265	0.6250	1.3130	29.000	35.000	1500.0	9300.0	0	Whimbrel
266	0.2500	0.4370	24.000	26.000	1250.0	5300.0	0	Spotted-Redshank
267	0.1250	0.2500	21.000	23.000	1850.0	7500.0	0	Marsh-Sandpiper
268	0.3130	0.6250	26.000	27.000	1500.0	7500.0	0	Common-Greenshank
269	0.1250	0.2500	22.000	24.000	300.0	5600.0	0	Green-Sandpiper
270	0.1090	0.2190	22.000	22.000	1500.0	8000.0	0	Wood-Sandpiper
271	0.1250	0.2500	22.000	23.000	1850.0	8000.0	0	Terek-Sandpiper
272	0.0790	0.1560	14.000	16.000	300.0	8000.0	0	Common-Sandpiper
273	2.1250	4.4370	58.000	66.000	155.0	3750.0	0	GreatBlackheaded-Gull
274	0.5630	0.8750	36.000	39.000	155.0	1850.0	0	Mediteranean-Gull
275	0.3750	0.7500	39.000	43.000	300.0	4700.0	0	Blackheaded-Gull
276	0.1880	0.3130	29.000	31.000	300.0	2175.0	0	Little-Gull
277	0.4370	0.6250	39.000	45.000	300.0	6000.0	0	Gullbilled-Tern
278	0.5000	0.6250	37.000	41.000	300.0	6800.0	0	Sandwich-Tern
279	0.2500	0.3750	30.000	38.000	600.0	8000.0	0	Common-Tern
280	0.2500	0.3130	29.000	32.000	300.0	1250.0	0	WhiteCheeked-Tern
281	0.1880	0.2500	29.000	30.000	1850.0	5000.0	0	Whiskered-Tern
282	0.1250	0.1880	24.000	26.000	2175.0	5600.0	0	WhiteWingedBlack-Tern
283	0.5630	0.8130	24.000	27.000	60.0	2500.0	0	Stock-Pigeon
284	0.3750	0.5000	23.000	24.000	300.0	1250.0	0	YellowEyedStock-Pigeon

(continued)

(continued)

Number	Mass range (kg/lb)		Span range (m/inch)		Migration range (km/mile)		Ind	Name
285	0.6880	1.3750	29.000	31.000	60.0	2175.0	0	CommonWood-Pigeon
286	0.3750	0.6250	20.000	23.000	300.0	2500.0	0	RufousTurtle-Dove
287	0.3130	0.4370	22.000	24.000	1850.0	2800.0	0	GreatSpotted-Cuckoo
288	0.1880	0.3130	20.000	22.000	300.0	7150.0	0	Oriental-Cuckoo
289	0.1250	0.2500	22.000	25.000	1850.0	7500.0	0	European-Nightjar
290	0.1250	0.1880	22.000	26.000	900.0	3400.0	0	Egyptian-Nightjar
291	0.0630	0.1090	16.000	18.000	1100.0	3400.0	0	Pallid-Swift
292	0.1880	0.2500	21.000	23.000	600.0	5600.0	0	Alpine-Swift
293	0.0940	0.1250	18.000	19.000	600.0	2500.0	0	Blue-Cheeked Be-eeater
294	0.0310	0.0630	9.000	11.000	300.0	2800.0	0	ShortToed-Lark
295	0.0470	0.0630	12.000	13.000	600.0	3000.0	0	RedRumped-Swallow
296	0.0310	0.0470	10.000	11.000	900.0	8000.0	0	House-Martin
297	0.0470	0.0630	11.000	12.000	900.0	3000.0	0	Blyths-Pipit
298	0.0310	0.0630	9.000	11.000	900.0	4475.0	0	Tawny-Pipit
299	0.0310	0.0630	9.000	10.000	300.0	5600.0	0	OliveTree-Pipit
300	0.0310	0.0630	9.000	10.000	600.0	6500.0	0	Tree-Pipit
301	0.0310	0.0630	9.000	10.000	5000.0	5600.0	0	Pechora-Pipit
302	0.0310	0.0470	8.000	9.000	60.0	3000.0	0	Meadow-Pipit
303	0.0310	0.0470	9.000	10.000	300.0	4700.0	0	Citrine-Wagtail
304	0.0310	0.0630	9.000	10.000	0.0	5300.0	0	Gray-Wagtail
305	0.0310	0.0630	9.000	11.000	0.0	5600.0	0	White-Wagtail
306	0.0310	0.0470	8.000	9.000	1850.0	2500.0	0	Siberian-Accentor
307	0.0470	0.0470	8.000	10.000	600.0	3400.0	0	RufousTailedScrub-Robin
308	0.0470	0.0630	9.000	10.000	2500.0	6500.0	0	Thrush-Nightingale
309	0.0310	0.0630	9.000	10.000	900.0	5000.0	0	Siberian-Rubythroat
310	0.0310	0.0310	7.000	8.000	1250.0	3400.0	0	SiberianBlue-Robin
311	0.0310	0.0310	8.000	9.000	600.0	3400.0	0	RedFlanked-Bluetail
312	0.0310	0.0630	10.000	11.000	1250.0	3400.0	0	WhiteThroated-Robin
313	0.0310	0.0470	8.000	9.000	1850.0	6000.0	0	Whinchat
314	0.0310	0.0470	7.000	8.000	0.0	5300.0	0	Common-Stonechat
315	0.0470	0.0790	10.000	12.000	300.0	3000.0	0	Isabelline-Wheatear
316	0.0310	0.0630	9.000	11.000	1850.0	5600.0	0	Pied-Wheatear
317	0.0310	0.0470	9.000	10.000	600.0	2175.0	0	BlackEared-Wheatear
318	0.0470	0.0630	10.000	12.000	300.0	1850.0	0	RedTailed-Wheatear
319	0.2500	0.4370	17.000	18.000	300.0	4000.0	0	Whites-Thrush
320	0.1560	0.2500	12.000	14.000	0.0	4000.0	0	Song-Thrush
321	0.2190	0.3130	16.000	18.000	0.0	3000.0	0	Mistle-Thrush

(continued)

(continued)

Number	Mass range (kg/lb)		Span range (m/inch)		Migration range (km/mile)		Ind	Name
322	0.0310	0.0470	6.000	7.000	2175.0	3750.0	0	PallasGrasshopper-Warbler
323	0.0310	0.0310	5.000	7.000	1850.0	5600.0	0	Lanceolated-Warbler
324	0.0310	0.0310	5.000	7.000	1850.0	4000.0	0	CommonGrass-hopper-Warbler
325	0.0310	0.0470	7.000	8.000	3750.0	5900.0	0	River-Warbler
326	0.0310	0.0470	7.000	8.000	2500.0	3700.0	0	Savis-Warbler
327	0.0310	0.0310	6.000	7.000	3000.0	4700.0	0	Aquatic-Warbler
328	0.0160	0.0470	6.000	8.000	2500.0	7150.0	0	Sedge-Warbler
329	0.0160	0.0310	5.000	7.000	600.0	4000.0	0	Paddyfield-Warbler
330	0.0160	0.0310	6.000	7.000	900.0	4000.0	0	BlythsReed-Warbler
331	0.0160	0.0310	7.000	8.000	3000.0	6500.0	0	Marsh-Warbler
332	0.0470	0.0940	9.000	11.000	1500.0	6500.0	0	GreatReed-Warbler
333	0.0470	0.0630	8.000	9.000	1850.0	3750.0	0	ThickBilled-Warbler
334	0.0160	0.0310	7.000	8.000	600.0	4350.0	0	Booted-Warbler
335	0.0310	0.0470	7.000	9.000	1500.0	3400.0	0	Upchers-Warbler
336	0.0310	0.0470	9.000	10.000	3000.0	4350.0	0	OliveTree-Warbler
337	0.0310	0.0310	7.000	9.000	3000.0	7500.0	0	Icterine-Warbler
338	0.0160	0.0310	6.000	7.000	1500.0	3400.0	0	Melodious-Warbler
339	0.0160	0.0310	5.000	7.000	1500.0	2500.0	0	Subalpine-Warbler
340	0.0160	0.0310	5.000	7.000	600.0	2800.0	0	Menetries-Warbler
341	0.0160	0.0310	7.000	8.000	1500.0	2175.0	0	Ruppells-Warbler
342	0.0160	0.0310	5.000	7.000	300.0	2500.0	0	Desert-Warbler
343	0.0310	0.0630	7.000	9.000	900.0	2800.0	0	Orphean-Warbler
344	0.0470	0.0630	9.000	10.000	1850.0	4700.0	0	Barred-Warbler
345	0.0160	0.0310	6.000	8.000	600.0	4000.0	0	Lesser-Whitethroat
346	0.0310	0.0790	7.000	9.000	2500.0	6800.0	0	Garden-Warbler
347	0.0310	0.0630	7.000	9.000	0.0	5900.0	0	Blackcap
348	0.0130	0.0260	6.000	7.000	2500.0	3000.0	0	Green-Warbler
349	0.0130	0.0260	5.000	8.000	600.0	4700.0	0	Greenish-Warbler
350	0.0130	0.0260	4.000	6.000	600.0	3400.0	0	PallasLeaf-Warbler
351	0.0130	0.0260	5.000	7.000	600.0	5600.0	0	YellowBrowed-Warbler
352	0.0160	0.0310	6.000	8.000	1250.0	2500.0	0	Raddes-Warbler
353	0.0160	0.0310	5.000	7.000	900.0	3000.0	0	Dusky-Warbler
354	0.0130	0.0260	6.000	7.000	1500.0	3000.0	0	Bonellis-Warbler
355	0.0130	0.0260	5.000	8.000	0.0	5600.0	0	Common-Chiffchaff
356	0.0160	0.0310	7.000	8.000	1850.0	5000.0	0	RedBreasted-Flycatcher
357	0.0310	0.0310	9.000	9.000	3000.0	3750.0	0	SemiCollared-Flycatcher
358	0.0160	0.0310	8.000	9.000	3000.0	6000.0	0	Collared-Flycatcher
359	0.1090	0.1880	17.000	18.000	600.0	6800.0	0	Golden-Oriole

(continued)

(continued)

Number	Mass range (kg/lb)		Span range (m/inch)		Migration range (km/mile)		Ind	Name
360	0.0940	0.1250	12.000	13.000	4000.0	6500.0	0	LesserGray-Shrike
361	0.0610	0.1090	10.000	11.000	1500.0	4000.0	0	Woodchat-Shrike
362	0.0310	0.0470	9.000	10.000	1500.0	2800.0	0	Masked-Shrike
363	3.3000	7.5000	1.800	2.100	2500.0	5000.0	1	Tundra-Swan
364	7.0000	12.5000	2.200	2.500	0.0	2500.0	1	Trumpeter-Swan
365	1.4000	2.2000	1.320	1.450	1800.0	3200.0	1	Barnacle-Goose
366	1.2000	1.5000	1.100	1.200	3000.0	6500.0	1	Brant
367	2.0000	3.0000	1.320	1.600	2000.0	5000.0	1	Lesser-SnowGoose
368	2.8000	4.2000	1.450	1.750	3500.0	5000.0	1	Greater-Snowgoose
369	0.2200	0.3000	0.810	0.850	10000.0	14000.0	1	Eskimo-Curlew
370	0.0900	0.1500	0.600	0.670	5000.0	13000.0	1	Pacific-GoldenPlover
371	0.0500	0.0800	0.460	0.580	4000.0	15000.0	1	Ruff
372	0.0950	0.2150	0.560	0.610	2500.0	16000.0	1	Red-Knot
373	0.0150	0.0250	0.200	0.230	2500.0	6000.0	1	Bluethroat
374	0.0350	0.0750	0.330	0.350	1000.0	6500.0	1	Redwing
375	0.0700	0.1250	0.390	0.420	1000.0	5000.0	1	Foeldfare
376	0.0220	0.0300	0.250	0.290	0.0	5000.0	1	Chaffinch
377	0.0300	0.0650	0.290	0.330	0.0	6000.0	1	Snow-Bunting
378	0.0300	0.0400	0.240	0.290	1000.0	6500.0	1	Lapland-Longspur
379	2.1880	4.1880	41.000	45.000	155.0	4700.0	0	RedThroated-Loon
380	2.8750	7.5000	43.000	51.000	155.0	4000.0	0	Arctic-Loon
381	6.1880	9.9370	50.000	57.000	60.0	4000.0	0	Common-Loon
382	9.3750	14.1250	53.000	59.000	300.0	3400.0	0	YellowBilled-Loon
383	3.9370	7.3750	53.000	66.000	500.0	2500.0	0	PinkFooted-Goose
384	3.1250	7.3750	51.000	64.000	1500.0	3400.0	0	GreaterWhite-Fronted-Goose
385	2.8750	5.5000	47.000	53.000	1850.0	3750.0	0	LesserWhite-Fronted-Goose
386	2.6250	3.9370	47.000	53.000	2300.0	2900.0	0	Ross-Goose
387	1.6250	3.0000	28.000	33.000	300.0	4000.0	0	Greater-Scaup
388	2.6870	4.4370	33.000	40.000	1500.0	9600.0	0	King-Eider
389	3.0630	4.0630	33.000	36.000	185.0	500.0	0	Spectacled-Eider
390	1.1250	2.1880	27.000	29.000	300.0	900.0	0	Stellars-Eider
391	1.5630	3.1880	31.000	35.000	300.0	4350.0	0	Black-Scoter
392	3.0000	4.3750	35.000	38.000	125.0	3000.0	0	WhiteWinged-Scoter
393	1.4370	2.5000	30.000	36.000	300.0	2800.0	0	Surf-Scoter
394	1.7500	4.6250	51.000	62.000	300.0	1500.0	0	GyrFalcon
395	0.1880	0.2500	22.000	25.000	1850.0	5600.0	0	Eurasian-Dotterel
396	0.3130	0.6250	26.000	29.000	300.0	4000.0	0	EuropeanGolden-Plover
397	0.3750	0.7500	27.000	32.000	1500.0	8700.0	0	BlackBellied-Plover
398	0.3130	0.4370	15.000	17.000	1500.0	9600.0	0	Sanderling

(continued)

(continued)

Number	Mass range (kg/lb)		Span range (m/inch)		Migration range (km/mile)		Ind	Name
399	0.0470	0.0940	13.000	14.000	4350.0	7150.0	0	Semipalmated-Sandpiper
400	0.0310	0.0790	13.000	14.000	4700.0	6800.0	0	Western-Sandpiper
401	0.0470	0.0940	13.000	14.000	3000.0	8700.0	0	RedNecked-Stint
402	0.0310	0.0940	13.000	14.000	2175.0	8400.0	0	Little-Stint
403	0.0310	0.0790	13.000	14.000	1250.0	5300.0	0	Temmincks-Stint
404	0.0470	0.0790	12.000	13.000	1500.0	5300.0	0	Least-Sandpiper
405	0.0470	0.0940	15.000	17.000	5300.0	9300.0	0	WhiteRumped-Sandpiper
406	0.0630	0.0940	15.000	18.000	5000.0	9600.0	0	Bairds-Sandpiper
407	0.0940	0.2340	16.000	19.000	5600.0	9600.0	0	Pectoral-Sandpiper
408	0.0940	0.2500	16.000	18.000	5600.0	9000.0	0	SharpTailed-Sandpiper
409	0.0790	0.2190	16.000	18.000	3750.0	9000.0	0	Curlew-Sandpiper
410	0.1250	0.2500	16.000	18.000	300.0	1850.0	0	Rock-Sandpiper
411	0.1250	0.2500	16.000	18.000	125.0	3750.0	0	Purple-Sandpiper
412	0.0940	0.1410	16.000	18.000	5300.0	6800.0	0	Stilt-Sandpiper
413	0.1090	0.2030	16.000	18.000	5600.0	6800.0	0	BuffBreasted-Sandpiper
414	0.2040	0.3130	18.000	20.000	3000.0	5000.0	0	LongBilled-Dowitcher
415	0.4370	1.0000	27.000	31.000	1250.0	9000.0	0	BarTailed-Godwit
416	0.3750	0.8750	25.000	29.000	5600.0	9300.0	0	Hudsonian-Godwit
417	0.4370	1.0000	28.000	32.000	5600.0	6000.0	0	BristleThighed-Curlew
418	0.1880	0.3130	16.000	20.000	2500.0	9300.0	0	Wandering-Tattler
419	0.1880	0.4370	19.000	22.000	300.0	9300.0	0	Ruddy-Turnstone
420	0.1250	0.3750	18.000	21.000	600.0	3750.0	0	Black-Turnstone
421	0.1880	0.3750	21.000	23.000	1250.0	9300.0	0	Surfbird
422	0.0630	0.1090	12.000	16.000	3000.0	6500.0	0	RedNecked-Phalarope
423	0.0790	0.1720	15.000	17.000	2500.0	3900.0	0	Red-Phalarope
424	1.1880	2.0000	49.000	54.000	600.0	9300.0	0	Pomarine-Jaeger
425	0.6880	1.4370	43.000	49.000	600.0	10000.0	0	Parasitic-Jaeger
426	0.3130	0.5000	35.000	39.000	1250.0	8700.0	0	Sabines-Gull
427	1.6250	1.8750	55.000	59.000	600.0	1850.0	0	Iceland-Gull
428	2.3750	4.0000	59.000	64.000	300.0	1850.0	0	Glaucous-Gull
429	1.6250	2.0000	55.000	59.000	1500.0	3000.0	0	Thayers-Gull
430	2.0000	3.5000	55.000	59.000	300.0	2800.0	0	GlaucousWinged-Gull
431	0.2500	0.5630	35.000	39.000	1500.0	3000.0	0	Ross-Gull
432	0.2190	0.2500	29.000	31.000	600.0	1850.0	0	Aleutian-Tern
433	1.6250	2.6880	25.000	28.000	900.0	1850.0	0	ThickBilled-Murre
434	0.6250	1.1250	20.000	22.000	125.0	1500.0	0	Black-Guillemot
435	0.4370	0.6250	15.000	18.000	300.0	1250.0	0	Crested-Auklet

(continued)

(continued)

Number	Mass range (kg/lb)		Span range (m/inch)		Migration range (km/mile)		Ind	Name
436	0.0630	0.1090	11.000	13.000	600.0	3000.0	0	Horned-Lark
437	0.0310	0.0470	9.000	10.000	3750.0	5900.0	0	RedThroated-Pipit
438	0.1250	0.1720	13.000	14.000	2500.0	4700.0	0	Siberian-Thrush
439	0.1410	0.1880	14.000	15.000	0.0	3750.0	0	Varied-Thrush
440	0.0470	0.0940	11.000	12.000	3000.0	6000.0	0	GrayCheeked-Thrush
441	0.0160	0.0310	6.000	8.000	1500.0	5900.0	0	Arctic-Warbler
442	0.0310	0.0470	6.000	8.000	300.0	1250.0	0	Hoary-Redpoll
443	0.0310	0.0630	7.000	10.000	1850.0	2800.0	0	Harris-Sparrow
444	0.0310	0.0630	6.000	9.000	1850.0	2800.0	0	Smiths-Longspur
445	0.0100	0.0150	0.220	0.250	0.0	8000.0	1	Blueandwhite-Swallow
446	0.0180	0.0250	0.230	0.270	0.0	5000.0	1	Darkfaced-GroundTyrant
447	0.0170	0.0200	0.190	0.210	0.0	800.0	1	Giant-Hummingbird
448	2.0000	2.6000	1.200	1.400	0.0	800.0	1	Kelp-Goose
449	0.0400	0.0600	0.450	0.550	0.0	3600.0	1	Twobanded-Plover
450	2.5000	3.5000	1.400	1.600	0.0	5.0	1	Andean-Goose
451	2.5000	3.5000	1.000	1.100	30.0	2500.0	1	Gray-Gull
452	3.3130	3.7500	39.000	42.000	125.0	900.0	0	Great-Greebe
453	2.4370	3.0630	39.000	43.000	300.0	2175.0	0	BlackFaced-Ibis
454	7.6870	14.7500	55.000	66.000	125.0	2175.0	0	BlackNecked-Swan
455	1.1250	1.3130	27.000	33.000	300.0	3000.0	0	Red-Shoveler
456	1.2500	1.3750	21.000	24.000	125.0	2175.0	0	ArgentineBlue-Billed-Duck
457	0.4370	0.6250	33.000	37.000	300.0	2500.0	0	Plumbeous-Kite
458	0.8750	1.4370	45.000	51.000	300.0	3000.0	0	LongWinged-Harrier
459	0.1250	0.1880	14.000	15.000	300.0	3750.0	0	Least-Seedsnipe
460	0.1880	0.2500	22.000	25.000	125.0	2000.0	0	RufousChested-Dotterel
461	0.0040	0.0040	5.000	6.000	125.0	300.0	0	GreenBacked-Firecrown
462	0.1250	0.1250	19.000	21.000	300.0	900.0	0	Striped-Cuckoo
463	0.0630	0.0630	9.000	11.000	300.0	1250.0	0	ForkTailed-Flycatcher
464	0.0630	0.0630	10.000	11.000	300.0	1500.0	0	Large-Elaenia
465	0.0470	0.0630	13.000	15.000	300.0	1500.0	0	BrownChested-Martin
466	0.0150	0.0150	0.160	0.160	1000.0	2000.0	1	AfricanPigmy-Flycatcher
467	2.0000	2.0000	1.100	1.400	3500.0	3900.0	1	African-Cmbduck
468	0.0600	0.0600	0.450	0.500	500.0	1200.0	1	Carmine-beeeater
469	0.0150	0.0150	0.240	0.280	500.0	1800.0	1	AfricanParadise-Flycatcher

(continued)

(continued)

Number	Mass range (kg/lb)		Span range (m/inch)		Migration range (km/mile)		Ind	Name
470	0.0210	0.0210	0.260	0.290	2000.0	2500.0	1	SouthAfricanCliff-Swallow
471	0.0210	0.0210	0.230	0.270	10.0	200.0	1	Starred-Robin
472	0.3000	0.3000	1.180	1.250	1000.0	1600.0	1	GreaterCrested-Tern
473	0.1880	0.2500	17.000	19.000	300.0	3000.0	0	Dwarf-Bittern
474	0.5000	0.5000	31.000	35.000	600.0	1850.0	0	MalagasyPond-Heron
475	2.6250	3.0630	43.000	51.000	600.0	4350.0	0	Abdims-Stork
476	1.1880	2.1880	29.000	33.000	1250.0	3000.0	0	Southern-Pochard
477	0.3750	0.4370	25.000	29.000	125.0	900.0	0	AfricanScissor-Tailed-Kite
478	0.9370	1.8750	49.000	55.000	60.0	600.0	0	Wahlbergs-Eagle
479	0.1880	0.2500	23.000	26.000	500.0	1100.0	0	Madagaskar-Pratincole
480	0.2190	0.2500	23.000	25.000	600.0	2500.0	0	BrownCVhested-Plover
481	0.0630	0.0940	19.000	20.000	1250.0	2800.0	0	Damara-Tern
482	0.1250	0.1880	17.000	19.000	600.0	2175.0	0	Jacobin-Cuckoo
483	0.1880	0.3130	21.000	23.000	300.0	1250.0	0	African-Cuckoo
484	0.0790	0.0940	13.000	15.000	300.0	600.0	0	AfricanEmarald-Cuckoo
485	0.0310	0.0940	12.000	13.000	300.0	1850.0	0	Didric-Cuckoo
486	0.1250	0.1880	25.000	26.000	1250.0	2500.0	0	PnnantWinged-Nightjar
487	0.0470	0.0630	13.000	14.000	300.0	2500.0	0	WhiteRumped-Swift
488	0.0790	0.1090	12.000	13.000	600.0	1500.0	0	GrayHeaded-Kingfisher
489	0.0790	0.1250	18.000	19.000	300.0	5000.0	0	BlueCheeked-Beeeater
490	0.0310	0.0470	12.000	13.000	300.0	900.0	0	RedBreasted-Swallow
491	0.0160	0.0310	6.000	7.000	300.0	1250.0	0	AfricanReed-Warbler
492	0.0630	0.0790	9.000	10.000	300.0	900.0	0	Amethyst-Starling
493	0.0500	0.0740	0.330	0.360	350.0	2500.0	1	Swift-Parrot
494	0.0280	0.0580	0.290	0.330	0.0	3900.0	1	Sacred-Kingfisher
495	0.0200	0.0330	0.400	0.450	0.0	4800.0	1	Australian-Beeeater
496	0.0120	0.0150	0.220	0.250	50.0	2500.0	1	Scarlet-Honeyeater
497	0.0750	0.2200	0.380	0.480	50.0	2400.0	1	Little-Friarbird
498	0.1500	0.2200	0.480	0.550	50.0	1600.0	1	Noisy-Friarbird
499	0.0150	0.0200	0.330	0.360	50.0	2000.0	1	Graybacked-Whiteeye
500	0.7500	0.8750	34.000	37.000	300.0	2800.0	0	Cattle-Egret
501	1.0000	1.5630	37.000	39.000	300.0	1250.0	0	Gray-Falcon
502	0.3130	0.6870	32.000	36.000	500.0	900.0	0	Australian-Hobby

(continued)

(continued)

Number	Mass range (kg/lb)		Span range (m/inch)		Migration range (km/mile)		Ind	Name
503	1.0000	1.1880	31.000	33.000	60.0	750.0	0	SouthIslandPied-Oystercatcher
504	0.1250	0.1560	20.000	22.000	500.0	870.0	0	Wrybill
505	0.1410	0.1720	21.000	23.000	300.0	2175.0	0	Australian-Pratincole
506	0.1250	0.1880	17.000	21.000	300.0	1500.0	0	DoubleBanded-Plover
507	0.1090	0.1250	11.000	13.000	155.0	470.0	0	OrangeBellied-Parrot
508	0.1250	0.1880	14.000	17.000	300.0	600.0	0	Cockatiel
509	1.3130	1.7500	51.000	62.000	300.0	2175.0	0	ChannelBilled-Cuckoo
510	0.1250	0.1410	17.000	20.000	1500.0	2175.0	0	LongTailed-Cuckoo
511	0.1250	0.1560	24.000	25.000	300.0	900.0	0	WhiteThroated-Nightjar
512	0.0470	0.0470	7.000	9.000	300.0	900.0	0	BuffBreasted-Paradise-Kingfisher
513	0.2500	0.3130	21.000	23.000	600.0	1850.0	0	Dollarbird
514	0.1250	0.1880	15.000	17.000	600.0	2500.0	0	RedBreasted-Pitta
515	0.0310	0.0470	10.000	11.000	600.0	3000.0	0	Tree-Martin
516	0.0940	0.1090	12.000	14.000	300.0	1250.0	0	Common-Cicadabird
517	0.0630	0.0790	10.000	11.000	600.0	2175.0	0	WhiteWinged-Tiller
518	0.0310	0.0310	8.000	9.000	60.0	300.0	0	Flame-Robin
519	0.0490	0.0630	8.000	9.000	300.0	900.0	0	AustralianReed-Warbler
520	0.0490	0.0630	7.000	9.000	300.0	900.0	0	Rufous-Songlark
521	0.0160	0.0310	6.000	8.000	125.0	900.0	0	WhiteThroated-Gerygone
522	0.0630	0.0790	9.000	11.000	300.0	1250.0	0	Blackfaced-Monarch
523	0.0630	0.0790	9.000	11.000	125.0	1100.0	0	Satin-Flycatcher
524	0.0310	0.0490	8.000	9.000	300.0	900.0	0	Rufous-Fantail
525	0.0310	0.0310	7.000	8.000	125.0	900.0	0	YellowFaced-Honeyeater
526	0.0310	0.0310	6.000	7.000	125.0	900.0	0	Eastern-Spinebill
527	6.2500	11.3000	2.500	3.500	5000.0	20000.0	1	Wandering-Albatross
528	3.0000	5.0000	2.300	2.600	5000.0	15000.0	1	Blackbrowed-albatross
529	2.1000	2.1000	2.400	2.400	1500.0	1500.0	1	Waved-Albatross
530	0.0340	0.0450	0.380	0.420	5000.0	15000.0	1	Wilson-StormPetrel
531	0.3500	0.5750	0.760	0.890	8500.0	12500.0	1	Manx-Shearwater
532	0.4800	0.8000	0.950	1.000	11000.0	13500.0	1	ShortTailed-Shearwater
533	0.2500	0.3600	1.050	1.170	8000.0	14400.0	1	LongTailed-Jaeger

(continued)

(continued)

Number	Mass range (kg/lb)		Span range (m/inch)		Migration range (km/mile)		Ind	Name
534	0.3000	0.5350	0.950	1.200	500.0	5000.0	1	Blacklegged-Kittiwake
535	0.5000	0.7000	1.080	1.200	200.0	2000.0	1	Ivory-Gull
536	0.0800	0.1200	0.750	0.850	15000.0	20000.0	1	Arctic-Tern
537	19.8130	21.0000	120.000	138.000	3750.0	9300.0	0	Royal-Albatross
538	5.3130	6.8130	80.000	83.000	600.0	7500.0	0	Bullers-Albatross
539	6.6250	7.9370	75.000	83.000	600.0	6000.0	0	Blackfooted-Albatross
540	5.0630	6.1880	76.000	79.000	600.0	4350.0	0	Lysan-Albatross
541	7.5000	9.6890	86.000	100.000	1250.0	7500.0	0	Shy-Albatross
542	5.5000	6.4370	78.000	100.000	600.0	7500.0	0	YellowNosed-Albatross
543	6.6250	8.2500	70.000	86.000	1250.0	9300.0	0	GrayHeaded-Albatross
544	5.3130	5.9370	78.000	80.000	1250.0	7500.0	0	Sooty-Albatross
545	6.1880	6.8130	72.000	85.000	1250.0	9300.0	0	LightMantledSooty-Albatross
546	8.3750	11.0000	70.000	78.000	600.0	9300.0	0	NorthernGiant-Petrel
547	1.6250	1.8750	44.000	47.000	600.0	3400.0	0	Southern-Fulmar
548	1.1250	1.6880	39.000	40.000	600.0	1500.0	0	Antarctic-Petrel
549	0.7500	1.0630	31.000	35.000	600.0	5000.0	0	Cape-Petrel
550	1.0000	1.6250	37.000	39.000	600.0	2800.0	0	GreatWinged-Petrel
551	1.2500	1.8130	42.000	43.000	600.0	6000.0	0	Whiteheaded-Petrel
552	0.8750	1.3130	35.000	39.000	600.0	1500.0	0	BlackCapped-Petrel
553	1.1250	1.5630	40.000	41.000	600.0	2800.0	0	Atlantic-Petrel
554	0.5630	1.0000	29.000	32.000	3000.0	8000.0	0	Mottled-Petrel
555	1.0000	1.1880	37.000	41.000	3000.0	5600.0	0	Solanders-Petrel
556	0.7500	0.8130	31.000	32.000	3000.0	7500.0	0	Kerguelen-Petrel
557	1.1250	1.1250	35.000	37.000	1250.0	3750.0	0	Kermadec-Petrel
558	1.0000	1.1880	37.000	38.000	2500.0	4700.0	0	JuanFernandez-Petrel
559	0.8130	1.1880	37.000	41.000	3000.0	5300.0	0	WhiteNecked-Petrel
560	0.3750	0.4370	25.000	25.000	3000.0	6500.0	0	Cooks-Petrel
561	0.3750	0.4370	25.000	26.000	1250.0	2500.0	0	MasATerra-Petrel
562	0.3130	0.4370	24.000	27.000	2800.0	5900.0	0	BlackWinged-Petrel
563	0.3130	0.3750	20.000	25.000	4350.0	6000.0	0	Steje=negers-Petrel
564	0.3750	0.5000	22.000	27.000	1850.0	5600.0	0	Blue-Petrel
565	0.3750	0.3750	22.000	22.000	600.0	4350.0	0	Salvins-Prion
566	0.3130	0.3750	22.000	25.000	600.0	3400.0	0	Antarctic-Prion
567	0.3130	0.3130	21.000	22.000	1500.0	4700.0	0	SlenderBilled-Prion
568	1.5000	1.5630	43.000	47.000	3750.0	6800.0	0	Black-Petrel
569	0.1880	0.3130	26.000	28.000	3000.0	5000.0	0	Bulwers-Petrel
570	2.0000	2.6870	45.000	51.000	1500.0	5600.0	0	Gray-Petrel

(continued)

(continued)

Number	Mass range (kg/lb)		Span range (m/inch)		Migration range (km/mile)		Ind	Name
571	2.2500	3.1250	52.000	57.000	1500.0	5600.0	0	WhiteChinned-Petrel
572	1.0000	1.1880	47.000	48.000	3750.0	5300.0	0	Streaked-Shearwater
573	1.2500	2.1250	39.000	49.000	2500.0	6000.0	0	Corys-Shearwater
574	1.5630	1.7500	42.000	43.000	3750.0	6800.0	0	PinkFooted-Shearwater
575	1.2500	1.6870	38.000	42.000	3750.0	8000.0	0	FleshFooted-Shearwater
576	1.5630	2.1250	39.000	46.000	4350.0	9000.0	0	Greater-Shearwater
577	0.7500	0.9370	37.000	38.000	4350.0	8700.0	0	Bullers-Shearwater
578	1.4370	2.1250	37.000	42.000	6000.0	8400.0	0	Sooty-Shearwater
579	0.7500	0.8130	28.000	35.000	1500.0	4700.0	0	Huttons-Shearwater
580	0.5000	0.9370	29.000	30.000	1850.0	2500.0	0	Fluttering-Shearwater
581	0.1250	0.1410	17.000	18.000	3000.0	5300.0	0	BlackBellied-StormPetrel
582	0.0790	0.0940	17.000	18.000	5000.0	8000.0	0	Swinhoes-StormPetrel
583	0.0790	0.0940	17.000	18.000	3000.0	5600.0	0	Leachs-StormPetrel
584	0.0940	0.1090	18.000	20.000	600.0	2800.0	0	Black-StormPetrel
585	0.1250	0.1250	21.000	22.000	4350.0	6000.0	0	Matsudairas-StormPetrel
586	2.6250	3.5000	51.000	55.000	3000.0	9300.0	0	SouthPolar-Skua
587	0.2500	0.3130	22.000	25.000	600.0	1500.0	0	WhiteCheeked_Tern
588	0.4370	0.6250	37.000	39.000	1250.0	3750.0	0	Cayenne-Tern
589	1.2000	2.9500	1.420	1.660	500.0	5000.0	1	Snowy-Owl
590	0.6000	1.6600	1.200	1.500	500.0	4500.0	1	RoughLegged-Hawk
591	0.0380	0.0420	0.250	0.280	500.0	3500.0	1	Red-Crossbill

Appendix D
Data on Bats (from Various Technical Sources)

Number	Mass range (kg)		Span range (m)		Area (m^2)	Speed (m/s)	Name
1	0.0110	0.0150	0.290	0.350	0.0	0.0	Brazilian-Free-Tailed
2	0.0070	0.0140	0.260	0.280	0.0	0.0	Evening
3	0.0080	0.0110	0.270	0.310	0.0	0.0	Silver-Haired
4	0.0140	0.0310	0.350	0.390	0.0	0.0	Northern-Yellow
5	0.0170	0.0350	0.380	0.390	0.0	0.0	Hoary
6	0.0090	0.0140	0.300	0.300	0.0	0.0	Seminole
7	0.0080	0.0140	0.260	0.330	0.0	0.0	Eastern-Red
8	0.0080	0.0140	0.260	0.300	0.0	0.0	Rafinesques-Bigeared
9	0.0110	0.0230	0.320	0.350	0.0	0.0	Big-Brown
10	0.0050	0.0080	0.210	0.260	0.0	0.0	Eastern-Pipistrelle
11	0.0030	0.0040	0.210	0.250	0.0	0.0	Eastern-Smallfooted
12	0.0070	0.0080	0.240	0.270	0.0	0.0	Idiana
13	0.0060	0.0070	0.220	0.260	0.0	0.0	Northern-Longeared
14	0.0080	0.0110	0.270	0.300	0.0	0.0	Gray
15	0.0050	0.0080	0.240	0.270	0.0	0.0	Southeastern
16	0.0070	0.0140	0.220	0.270	0.0	0.0	Little-Brown
17	1.5000	1.5000	1.800	1.800	0.0	0.0	Flying-Fox
18	0.0140	0.0140	0.150	0.150	0.0	0.0	Kittis-Hognosed
19	0.1040	0.1040	0.530	0.530	0.0	0.0	Rousettus-Aegyptiacus
20	0.0526	0.0526	0.401	0.401	0.0	0.0	Epomophorus-Anurus
21	0.0053	0.0053	0.209	0.209	0.0	0.0	Pipistrellus-Pipistrellus
22	0.0265	0.0265	0.344	0.344	0.0	0.0	Nyctalus-Noctula
23	0.0099	0.0099	0.277	0.277	0.0	0.0	Eptesious-Nilssoni
24	0.0141	0.0141	0.298	0.298	0.0	0.0	Vespertilo-Murinus
25	0.0090	0.0090	0.270	0.270	0.0	0.0	Plecotus-Auritus
26	0.0356	0.0356	0.449	0.449	0.0	0.0	Otomops-Martiensseni
27	0.0130	0.0130	0.293	0.293	0.0	6.7	Lasiurus-Borealis
28	0.0310	0.0310	0.356	0.356	0.0	7.7	Lasiurus-Cinereus
29	0.3000	0.3000	0.750	0.750	0.0	0.0	Strawcolored-Fruit

(continued)

L. Kantha, *Migration on Wings*, SpringerBriefs in Applied Sciences and Technology,
DOI: 10.1007/978-3-642-27925-6, © The Author(s) 2012

(continued)

Number	Mass range (kg)		Span range (m)		Area (m²)	Speed (m/s)	Name
30	0.1300	0.1300	0.600	0.600	0.0	0.0	Egyptian-Fruit
31	0.1000	0.1000	0.400	0.400	0.0	0.0	Bocages-Fruit
32	0.0700	0.1100	0.500	0.500	0.0	0.0	Wahlbergs-Epaulated-Fruit
33	0.0800	0.1400	0.560	0.560	0.0	0.0	Peters-Epaulated-Fruit
34	0.0280	0.0280	0.340	0.340	0.0	0.0	Mauritian-Tomb
35	0.1200	0.1200	0.600	0.600	0.0	0.0	Commersns-Leafnosed
36	0.0080	0.0080	0.200	0.200	0.0	0.0	Sundevalls-Leafnosed
37	0.0050	0.0050	0.150	0.150	0.0	0.0	Shorteared-Trident
38	0.0120	0.0120	0.350	0.350	0.0	0.0	Persian-Leafnosed
39	0.0400	0.0400	0.350	0.350	0.0	0.0	Large-Slitfaced
40	0.0110	0.0110	0.240	0.240	0.0	0.0	Egyptian-Slitfaced
41	0.0270	0.0270	0.390	0.390	0.0	0.0	Hildebrandts-Horseshoe
42	0.0170	0.0170	0.320	0.320	0.0	0.0	Geoffroys-Horseshoe
43	0.0060	0.0060	0.200	0.200	0.0	0.0	Dents-Horseshoe
44	0.0100	0.0100	0.280	0.280	0.0	0.0	Schreibers-Longfingered
45	0.0065	0.0065	0.240	0.240	0.0	0.0	Cape-Serotine
46	0.0110	0.0110	0.280	0.280	0.0	0.0	Temmincks-Hairy
47	0.0040	0.0040	0.190	0.190	0.0	0.0	Banana
48	0.0130	0.0130	0.280	0.280	0.0	0.0	Butterfly
49	0.0045	0.0045	0.180	0.180	0.0	0.0	Schlieffens
50	0.0270	0.0270	0.300	0.300	0.0	0.0	Yellow-House
51	0.0160	0.0160	0.280	0.280	0.0	0.0	LesserYellow-House
52	0.0080	0.0080	0.250	0.250	0.0	0.0	Damar-Woolly
53	0.0130	0.0130	0.260	0.260	0.0	0.0	Flatheaded-Flattailed
54	0.0150	0.0150	0.300	0.300	0.0	0.0	Egyptian-Freetailed
55	0.0500	0.0500	0.450	0.450	0.0	0.0	Midas-Freetailed
56	0.0110	0.0110	0.240	0.240	0.0	0.0	Little-freetailed
57	0.0150	0.0500	0.160	0.250	0.0	0.0	Common-Vampire
58	0.0090	0.0090	0.27	0.270	0.0	0.0	Long-eared

Appendix E
Data on Insects from Technical Publications

Number	Mass (kg)	Area (m²)	Span (m)	Speed (m/s)	Name
1	0.276E-03	0.184E-03	0.313E-01	4.00	Tabanus-borinus-diptera
2	0.450E-04	0.360E-04	0.140E-01	2.00	Sarcophaga-carnariaL-diptera
3	0.120E-04	0.200E-04	0.110E-01	2.00	Musca-domestica-diptera'
4	0.730E-04	0.780E-04	0.240E-01	3.50	Volucella-pellucensMeig-diptera
5	0.180E-03	0.574E-04	0.202E-01	2.05	Tabanus-affinis(Horsefly)-diptera
6	0.350E-05	0.360E-05	0.538E-02	1.00	Aedes-nearcticus(Mosquito)-diptera
7	0.614E-03	0.172E-03	0.360E-01	4.00	Xylocopa-violacea-hymenoptera
8	0.388E-03	0.142E-03	0.320E-01	3.00	Bombus-terrestrisFabr-hymenoptera
9	0.187E-03	0.980E-04	0.280E-01	2.50	Vespa-germanica-hymenoptera
10	0.567E-03	0.260E-03	0.450E-01	6.00	Vespa-crabroL-hymenoptera
11	0.780E-04	0.420E-04	0.170E-01	2.50	Apis-mellificaL-hymenoptera
12	0.450E-04	0.420E-04	0.180E-01	1.50	Amonophila-sabulosaVdel-hymenoptera
13	0.300E-03	0.360E-02	0.740E-01	3.50	Papilo-podalirius-lepidoptera
14	0.134E-03	0.108E-02	0.540E-01	4.00	Vanesa-atalantaL-lepidoptera
15	0.127E-03	0.184E-02	0.620E-01	2.50	Pieris-brassicaL-lepidoptera
16	0.345E-03	0.400E-03	0.400E-01	5.00	Macroglossa-stellatorumL-lepidoptera
17	0.144E-03	0.440E-03	0.360E-01	1.50	Plusia-gammaL-dioptera
18	0.961E-03	0.402E-03	0.560E-01	2.50	Melolontha-vlgaris(Beetle)-coleoptera
19	0.961E-03	0.104E-02	0.902E-01	2.50	Melolontha-vlgaris(Beetle)-coleoptera
20	0.537E-03	0.260E-03	0.400E-01	3.00	Cetonia-aurata-coleoptera
21	0.537E-03	0.630E-03	0.622E-01	3.00	Cetonia-aurata-coleoptera
22	0.260E-02	0.800E-03	0.720E-01	1.50	Lucanus-corvus-coleoptera
23	0.260E-02	0.202E-02	0.114E+00	1.50	Lucanus-corvus-coleoptera
24	0.109E-03	0.116E-03	0.250E-01	0.80	Telephorus-fuscus-coleoptera
25	0.109E-03	0.282E-03	0.390E-01	0.80	Telephorus-fuscus-coleoptera
26	0.557E-03	0.120E-02	0.730E-01	5.00	Brachytron-pratenseMull-neuroptera
27	0.120E-03	0.850E-03	0.600E-01	1.50	Caopteryx-splendensHarr-neuroptera

(continued)

L. Kantha, *Migration on Wings*, SpringerBriefs in Applied Sciences and Technology, DOI: 10.1007/978-3-642-27925-6, © The Author(s) 2012

(continued)

Number	Mass (kg)	Area (m²)	Span (m)	Speed (m/s)	Name
28	0.380E-04	0.355E-03	0.500E-01	0.60	Pyrosoma-minimumHarr-neuroptera
29	0.300E-04	0.176E-03	0.290E-01	0.50	Panorpa-communisL-neuroptera
30	0.248E-03	0.108E-02	0.650E-01	4.00	Orthetrum-caerulescensFabr-neuroptera
31	0.530E-03	0.138E-02	0.790E-01	7.00	Aeschna-mixtraLatr-neuroptera
32	0.200E-02	0.266E-02	0.120E+00	3.50	Schistocera-gregariaLocust-orthoptera
33	0.790E-03	0.220E-02	0.984E-01	7.20	Anax-parthenope-odonata
34	0.330E-07	0.134E-05	0.000E+00	0.00	Besmisia-tabaci-homoptera
35	0.650E-07	0.194E-04	0.000E+00	0.00	Aleurothrixus-floccosus-homoptera
36	0.114E-06	0.103E-05	0.000E+00	0.00	Aphis-gossypii-homoptera
37	0.702E-06	0.111E-04	0.000E+00	0.00	Acryrthosiphon-kondoi-homoptera
38	0.212E-03	0.597E-02	0.143E+00	1.60	Alsomitra-macrocarpa
39	0.600E-02	0.300E-02	0.800E-01	0.00	Swallowtail-butterfly
40	0.120E-01	0.270E-02	0.800E-01	0.00	Blue-underwing
41	0.112E-01	0	0.100E+00	0.00	Moth
42	0.880E-02	0	0.346E-01	0.00	Bumblebee
43	0.100E-01	0.400E-03	0.600E-01	0.00	Cock-chafer
44	0.250E-06	0.250E-06	0.140E-02	0.00	Chalcid-Wasp
45	0.700E-05	0.210E-05	0.350E-02	0.00	fruit-fly
46	0.280E-03	0	0.346E-01	0.00	Crane-fly
48	0.150E-02	0	0.254E-01	0.00	Hover-fly
49	0.200E-04	0	0.600E-02	0.00	Fruit-fly
50	0.230E+00	0	0.730E+00	0.00	Meganeura-Dragonfly(LD ∼ 14)

Appendix F
Data from Pennycuick (1989) (Reproduced with Permission)

Number	Mass (kg)	Span (m)	Wing area (m^2)	Name
1	5.38	2.18	0.69	Vulture
2	5.19	1.99	0.33	Giant-Petrel
3	5.50	2.40	0.72	Crane
4	0.0174	0.23	0.01	Ovenbird
5	0.0103	0.167	0.0	Yellow-Throat
6	0.032	0.285	0.0	Swansons-Thrush
7	0.0837	0.40	0.0	American-Robin
8	0.0358	0.286	0.0	Cedal-Waxwing
9	0.0446	0.396	0.0	Yellowbellied-Sapsucker
10	0.288	1.00	0.0	Laughing-Gull
11	0.102	0.43	0.0	Mourning-Dove
12	8.73	3.03	0.0	Wandering-Albatross
13	1.37	1.40	0.0	Whitechinned-Petrel
14	0.433	0.875	0.0	Cape-Pigeon
15	0.038	0.393	0.0	Wilsons-StormPetrel
16	0.195	0.875	0.0	Sooty-Tern
17	3.30	2.39	0.0	Brown-Pelican
18	1.15	1.51	0.0	Brown-Booby
19	0.362	0.95	0.0	Whitetailed-TropicBird
20	1.42	1.16	0.0	DoubleCrested-Cormorant
21	1.49	2.25	0.0	Magnificent-FrigateBird
22	1.87	1.76	0.0	Great-Blue-Heron
23	0.97	0.99	0.0	White-Ibis
24	2.22	1.80	0.0	Turkey-Vulture
25	4.20	2.18	0.0	Bald-Eagle
26	9.50	2.20	0.0	Tundra-Swan
27	7.57	2.41	0.0	RipellsGriffon-Vulture

L. Kantha, *Migration on Wings*, SpringerBriefs in Applied Sciences and Technology,
DOI: 10.1007/978-3-642-27925-6, © The Author(s) 2012

Appendix G
Data from Technical Publications

Number	Mass (kg)	Span (m)	Speed (m/s)	Name
1	8.5000	3.4300	0.0	Albatross
2	0.6000	0.7000	0.0	Wood-Duck
3	0.8000	0.8500	0.0	Wood-Duck
4	3.3000	1.6000	0.0	Canada-Goose
5	0.0174	0.3300	11.1	Storm-Petrel
6	0.0180	0.3300	8.9	Swallow
7	0.0050	0.1500	11.9	Hummingbird
8	0.0330	0.3800	6.4	Swift
9	0.0300	0.2500	9.7	House-Sparrow
10	0.0800	0.3900	9.4	Starling
11	0.2000	0.7000	8.9	Kestrel
12	1.3200	1.4300	11.9	Heron
13	1.3500	1.6100	11.9	Heron
14	1.1800	1.4300	11.4	Herring-Gull
15	4.0000	2.0000	18.6	White-Stork
16	0.4700	0.8900	13.9	Carrion-Crow
17	8.5000	3.4100	15.0	Wang-Albatross
18	1.1100	0.9000	18.1	Mallard
19	0.0200	0.3400	0.0	Large-hummingbird-PatagonaGigas
20	0.2000	0.5800	0.0	Archaeopteryx
21	1.9000	2.7300	0.0	Nyctosaurus
22	15.0000	7.0000	0.0	Pteranodon
23	65.0000	11.0000	0.0	Pterosaur-Quetzalcoatlus-Northropi

L. Kantha, *Migration on Wings*, SpringerBriefs in Applied Sciences and Technology,   79
DOI: 10.1007/978-3-642-27925-6, © The Author(s) 2012

Appendix H
Bird Data from "Winged Migration by J. Perrin, 2003" (Reproduced with Permission)

Number	Mass range (kg)		Span range (kg)		Speed range (km)		Name
1	1.32	2.00	1.29	1.45	16.9	22.0	Barnacle-goose
2	2.50	4.40	1.47	1.78	16.3	22.0	Greylag-goose
3	5.08	6.08	2.18	2.44	12.3	18.9	Common-crane
4	0.10	0.15	0.76	0.84	11.0	11.0	Arctic-tern
5	9.07	15.00	2.31	3.56	8.4	13.6	White-Pelican
6	2.31	4.40	1.73	1.93	12.3	12.3	White-Stork
7	3.00	6.35	1.68	2.41	8.4	13.6	Bald-Eagle
8	3.00	5.00	1.30	1.60	15.0	26.0	Snow-goose
9	2.00	7.00	1.19	1.88	18.9	24.6	Canada-Goose
10	1.00	1.50	1.14	1.35	22.0	27.3	Redbreasred-Goose
11	2.00	3.18	1.40	1.57	16.3	22.0	Barheaded-Goose
12	1.50	1.50	1.09	1.19	8.8	13.2	Blackheaded-Ibis
13	6.81	11.80	2.18	2.46	13.2	22.0	Japanese-Crane
14	7.71	13.20	2.13	2.44	15.4	22.0	Whooper-Swan
15	5.90	10.90	2.54	3.43	11.0	22.0	Wandering-Albatross

L. Kantha, *Migration on Wings*, SpringerBriefs in Applied Sciences and Technology,
DOI: 10.1007/978-3-642-27925-6, © The Author(s) 2012

L. Kantha, *Migration on Wings*, SpringerBriefs in Applied Sciences and Technology,   83
DOI: 10.1007/978-3-642-27925-6, © The Author(s) 2012